GAS TURBINE PROPULSION SYSTEMS

Aerospace Series List

Advanced Control of Aircraft, Rockets and Spacecraft	Tewari	July 2011
Basic Helicopter Aerodynamics: Third Edition	Seddon *et al.*	July 2011
Cooperative Path Planning of Unmanned Aerial Vehicles	Tsourdos *et al.*	November 2010
Principles of Flight for Pilots	Swatton	October 2010
Air Travel and Health: A Systems Perspective	Seabridge *et al.*	September 2010
Design and Analysis of Composite Structures: With applications to aerospace Structures	Kassapoglou	September 2010
Unmanned Aircraft Systems: UAVS Design, Development and Deployment	Austin	April 2010
Introduction to Antenna Placement & Installations	Macnamara	April 2010
Principles of Flight Simulation	Allerton	October 2009
Aircraft Fuel Systems	Langton *et al.*	May 2009
The Global Airline Industry	Belobaba	April 2009
Computational Modelling and Simulation of Aircraft and the Environment: Volume 1 - Platform Kinematics and Synthetic Environment	Diston	April 2009
Handbook of Space Technology	Ley, Wittmann Hallmann	April 2009
Aircraft Performance Theory and Practice for Pilots	Swatton	August 2008
Surrogate Modelling in Engineering Design: A Practical Guide	Forrester, Sobester, Keane	August 2008
Aircraft Systems, 3rd Edition	Moir & Seabridge	March 2008
Introduction to Aircraft Aeroelasticity And Loads	Wright & Cooper	December 2007
Stability and Control of Aircraft Systems	Langton	September 2006
Military Avionics Systems	Moir & Seabridge	February 2006
Design and Development of Aircraft Systems	Moir & Seabridge	June 2004
Aircraft Loading and Structural Layout	Howe	May 2004
Aircraft Display Systems	Jukes	December 2003
Civil Avionics Systems	Moir & Seabridge	December 2002

GAS TURBINE PROPULSION SYSTEMS

Bernie MacIsaac
Retired Founder and CEO, GasTOPS Ltd, Canada

Roy Langton
Retired Group VP Engineering, Parker Aerospace, USA

A John Wiley & Sons, Ltd., Publication

Library of Congress Cataloguing-in-Publication Data

MacIsaac, Bernie.
    Gas turbine propulsion systems / Bernie MacIsaac, Roy Langton.
        p. cm.
    Includes bibliographical references and index.
    ISBN 978-0-470-06563-1 (hardback)
 1. Airplanes – Turbojet engines. 2. Jet boat engines. 3. Vehicles, Military. I. Langton, Roy. II. Title.
    TL709.3.T83M25 2011
    629.134′353–dc23

                                                                            2011016566

A catalogue record for this book is available from the British Library.

Print ISBN: 978-0-470-06563-1
ePDF ISBN: 978-1-119-97549-6
oBook ISBN: 978-1-119-97548-9
ePub ISBN: 978-1-119-97614-1
Mobi ISBN: 978-1-119-97615-8

Typeset in 10/12pt Times Roman by Laserwords Private Limited, Chennai, India

FSC
www.fsc.org
MIX
Paper from
responsible sources
FSC® C013604

Contents

About the Authors

BD (Bernie) MacIsaac

Dr MacIsaac received an Honors B. Eng. (Mechanical) from the Technical University of Nova Scotia in 1970. He was awarded a Science '67 graduate scholarship which took him to Ottawa to study Jet Engine Dynamics and Controls at Carleton University. He was awarded an M.Eng. in 1972 and a Ph.D. in 1974.

Following completion of his studies, Dr MacIsaac spent four years at the National Research Council of Canada where he helped to develop the first 8-bit microprocessor control for general aviation gas turbines. He was awarded a patent for a control design to prevent in-flight engine stalls on helicopter engines.

Dr MacIsaac formed GasTOPS Ltd. (Gas Turbines and Other Propulsion Systems) in 1979, an Ottawa-based company which specializes in the application of intelligent systems to machinery protection and machinery maintenance systems. Much of this company's work has focused on aerospace and industrial power plants. About 1991, GasTOPS began the development of an on-line oil debris detector for damage recognition of the oil-wetted components of power plants. This device is now fitted to many modern fighter aircraft, many land-based CoGen and pipeline engines and is selling well in the new emerging wind turbine market. This development has led to the establishment of a manufacturing facility and to worldwide sales of this product.

Dr MacIsaac served as GasTOPS Ltd. President until 2007, at which point he turned management of the company over to his longtime colleague Mr David Muir. Since then, Dr MacIsaac has devoted his efforts to the establishment of an R&D group at GasTOPS, which is responsible for the definition and subsequent demonstration of new technologies that will form the basis of the next product line for GasTOPS.

Dr MacIsaac participates as a lecturer in professional practice courses at both Ottawa and Carleton Universities as well as Carleton University-sponsored short courses on gas turbines.

He is a past president of the Canadian Aeronautics and Space Institute and is a past Chairman of PRECARN, a network of companies engaged in collaborative applied research. He currently serves as Chairman of the Senior Awards Committee of the Canadian Aeronautics and Space Institute.

Dr MacIsaac was born in 1945. He is married (1969) and has twin daughters who were born on Christmas Day in 1973 and three granddaughters and one grandson. He has lived in Ottawa, Canada with his wife Ann since 1970.

Roy Langton

Roy Langton began his career as a Student Apprentice in 1956 with English Electric Aviation (now BAE Systems) at Warton in Lancashire, UK. After graduating in Mechanical and Aeronautical Engineering, he worked on powered flight control actuation systems for several military aircraft, including the English Electric Lightning, the Anglo-French Jaguar, and Panavia Tornado.

In 1968 he emigrated to the USA working for Chandler Evans Corporation in West Hartford Connecticut (now part of the Goodrich Corporation) and later with Hamilton Standard (now Hamilton Sundstrand) on engine fuel controls as the technology transitioned from hydromechanics to digital electronics. During this period he was exposed to a wide variety of projects from small gas turbines such as the Tomahawk Missile cruise-engine to large, high-bypass gas turbines used on today's commercial transports. A major milestone during this period was the introduction of the first FADEC into commercial service on the Pratt & Whitney PW2037 engine, which powers many of the Boeing 757 aircraft.

In 1984, he began a career in aircraft fuel controls with Parker Hannifin Corporation as Chief Engineer for the Fuel Products Division of the Corporation's Aerospace Group in Irvine California. In the 20-year period prior to his retirement in 2004 as Group Vice-President of Engineering, he played a major role in establishing Parker Aerospace as a leading supplier of complete fuel systems to aircraft manufacturers around the world. This began in 1993 with the Bombardier Global Express business jet and culminated in 2000 with the Fuel Measurement and Management system for the A380 superjumbo commercial transport.

Roy Langton was born in 1939, married his wife June in 1960 and has two daughters and five grandchildren. Roy and June currently reside in Boise Idaho USA.

Roy continues to work as a part-time technical consultant for Parker Aerospace and has been an Aerospace Series Editor for John Wiley & Sons since 2005.

Preface

The gas turbine industry began in the 1940s and, for many decades, it remained an object of research by universities and government laboratories as well as the many commercial establishments which sprang to life in an effort to exploit the technology. During this period, much basic research was conducted and information exchange was encouraged. It is noteworthy that the British Government, which had sponsored much of the development of the Whittle engine, shared the entire technical package with the US Government as a war measure. This resulted in the US Government supporting its continued development at the General Electric facilities at Lynn, Massachusetts.

Many companies were formed in Europe and in North America during the 1950s, each of which offered designs tailored to specific applications. In addition to the rapidly expanding aeronautical and defense industries, other applications began to emerge for non-aeronautical engines. These included gas pipelines, electrical power generation and naval propulsion. In short, the industry was booming and employment for engineers was readily obtained. More importantly, there were many opportunities to learn about this fascinating machine.

Today, the industry is reduced to a handful of very large companies. The investment required to develop an engine is enormous and the competition can only be described as fierce. Engineers are much more specialized and commercial secrecy is a fundamental element of corporate survival. For the true engineering specialist, the work remains a fascinating push into the unknown. For the systems engineer who must develop strategies and equipment which supports and manages the operation of the engine, the work has however become more complex and information has become more difficult to obtain in a form that allows synthesis of system behavior.

There are many books available that describe gas turbine engines, focusing primarily on the 'turn and burn' machinery from an aerothermodynamic perspective. Typically, the coverage given to the peripheral systems that support the complete gas turbine propulsion system is either not described at all or is often superficial. As the industry continues to demand improvements in performance and reductions in weight, the engine continues to be refined and, in some instances, made more complex. The system engineer can therefore expect to be working on not only more refined control systems but also information management systems designed to keep ownership costs as low as possible.

This book is organized to provide the reader with a basic understanding of how a gas turbine works, with emphasis on those aspects of its operation which most affect the task of the system designer. We have attempted to cover the propulsion package as a combination of functional components that must operate properly in unison to produce power. The famous remark by Sir Frank Whittle–that the gas turbine has only one moving part–happily neglects the many subsystems that must operate in unison with the prime mover to create a viable propulsion system package. In Whittle's day, it was sufficient for the engine to run smoothly. Today, the complete engine design must take into account cost of ownership, maintainability, safety, and prognostics and health monitoring.

The book describes the basic gas turbine in terms of its major components at a level sufficient to understand its operation and to appreciate the hard limits of its operating envelope. In particular, the issues associated with the handling of the gas generator or 'core' of the turbine engine in aircraft propulsion applications in preventing the onset of compressor surge or flame-out during transient throttle changes is addressed in some depth, including the need for stable speed governing in steady-state operation.

The importance of understanding and managing the engine inlet and exhaust systems together with the issues associated with power extraction and bearing lubrication are also given extensive coverage.

The gas turbine has found application in a number of important non-aeronautical industries. These include pipeline compressor drives, electrical power generation and naval propulsion systems. From a systems design perspective, the naval application is arguably the most demanding. In keeping with the propulsion focus of this book, the naval application has been chosen as an example of the challenges of introducing the gas turbine engine–developed for airborne applications–into such a hostile environment. The subsystems required to support and protect the engine in a navy ship are described in some detail.

Finally, prognostics and health monitoring must be recognized as a key aspect of the need to develop reliable algorithms that can effectively forecast the operational life remaining. This is increasingly important as both the commercial and military operators move into the realm of condition-based maintenance as a means of controlling and minimizing cost of ownership. Some of these systems will be fitted to future engines; as their underlying advantages are recognized, it is of equal importance that they interact with ground-based logistics systems.

Notwithstanding the book's focus on the system aspects of gas turbine propulsion systems, the fundamentals of gas turbine engine design are covered to a level that is considered more than adequate for the practicing systems engineer and/or business program manager. In addition to the devotion of one complete chapter to gas turbine basics, there are several appendices that provide a substantial grounding in the fundamentals of gas turbine design, modeling and operation.

Series Preface

The propulsion system of an aircraft performs a number of key functions. Firstly it provides the propulsive energy to propel the aircraft throughput its route or mission with the appropriate achievement of performance, efficiency, safety and availability. Secondly it provides the prime source of energy for the on-board systems by enabling the generation of electrical, hydraulic and pneumatic power for their effectors. Finally it provides the air to create a habitable environment for crew, passengers and avionic equipment. It is also a major capital item in any modern commercial and military aircraft and its incorporation into the aircraft affects both airframe and systems, not only in technical interface terms, but also in terms of safety, reliability and cost of ownership.

Unsurprisingly then, a knowledge of the propulsion system is key to understanding how to integrate it with the airframe and the aircraft systems. Other books in the *Aerospace Series* cover the topics of aircraft performance, avionic and aircraft systems – all of which depend on the propulsion system to complete their tasks. A number of these systems have an intimate link with the propulsion system such as aerodynamics, structural design, fuel types, onboard fuel storage and system design, cabin environment and cooling, hydraulic and electrical generation, flight control, flight management, flight deck displays and controls, prognostic and health management, and finally systems modelling. The degree of integration of these systems varies with aircraft role and type, but in all cases the design of the systems cannot be complete without an understanding of the system that provides their energy.

This book, *Gas Turbine Propulsion Systems*, provides the key to that understanding by describing the propulsion system in terms of its major sub-systems with a suitable and readily understandable treatment of the underlying mathematics. An important point is that the book completes the picture of the aircraft systems by taking a systems engineering approach to propulsion. It deals, not only with the gas turbine engine and its aero-thermodynamics, but with the propulsion system as an integrated set of sub-systems that control the engine throughout the flight envelope and provide suitable controlled off-takes. The treatment of fuel control, thrust control, installation aspects and prognostics clearly link into integration of the propulsion system with the aircraft and its systems for pure gas turbines and shaft power turbines.

For good measure there is a chapter devoted to marine propulsion systems, and appendices complete the treatment of the underlying theory and provide guidance on thermodynamic modelling. There is also a discussion of the future direction of propulsion systems that addresses some aspects of reducing engine off-takes and contributes to the more-electric aircraft concept.

This is a book for all practising aircraft systems engineers who want to understand the interactions between their systems and the provider of their power source.

Allan Seabridge, Roy Langton, Jonathan Cooper and Peter Belobaba

Acknowledgements

This book has been completed with the help of many colleagues and organizations who were able to provide valuable information and support, specifically:

- Herb Saravanamuttoo of Carlton University;
- Richard Dupuis, Peter MacGillivray, Shawn Horning, and Doug Dubowski of GasTOPS; and
- Jean-Pierre Beauregard of Pratt & Whitney Canada (retired).

In particular, the authors would like to acknowledge the support received on three specific topics:

1. the Pratt & Whitney Canada PW150A engine control system;
2. the Concorde air inlet control system; and
3. the Meggitt Engine Monitoring Unit installed on all of the A380 engine options.

The first subject, addressed in Chapter 5, describes a modern turboprop application embodying a state-of-the art FADEC-based control system. In support of this topic, the authors would like to thank Pratt & Whitney Canada and particularly Jim Jarvo for his consultant services and active participation in the generation and review of the material. Jim is currently a Control Systems Fellow in the Engineering department of Pratt Whitney Canada based in Longueil, Quebec, Canada.

Regarding the second topic, the authors would like to thank the British Aircraft Corporation (now BAE Systems) for access to historical technical documents describing the Concorde air inlet system. We would also like to thank Roger Taplin who was the Lead Engineer on the Concorde AICS project during the design, development, and operational launch phases of the program. Roger, who is currently employed by Airbus at their Filton (UK) facility in the position of Aircraft Architect-Wing, provided valuable consultant and editorial support throughout the generation of the material presented in Chapter 6.

Thirdly, the authors are grateful for the information and support provided by Mervyn Floyd of Meggitt Plc in the UK concerning one of their most recent Engine Monitoring Unit programs. This topic is covered in Chapter 10 in support of the prognostics and health monitoring discussion.

In addition, the authors would like to acknowledge the following organizations that provided an important source of information through published material in support of the preparation of this book:

- Boeing;
- CFM International;
- General Electric Honeywell;
- Parker Aerospace;
- Pratt & Whitney; and
- Rolls-Royce.

List of Acronyms

ACARS	Aircraft Communication And Reporting System
ADC	Air Data Computer
AFDX	Avionics Full Duplex Switched Ethernet
AICS	Air Inlet Control System
AICU	Air Inlet Control Unit
AMAD	Aircraft Mounted Accessory Drive
APU	Auxiliary Power Unit
ARINC	Aeronautical Radio Incorporated
ASM	Air Separation Module
C-D	Convergent-Divergent
CDP	Compressor Delivery Pressure
CDU	Cockpit Display Unit
CFD	Computer Fluid Dynamics
CLA	Condition Lever Angle
CMC	Ceramic-Metal Composite
CPP	Controllable Pitch Propeller
CRP	Controllable Reversible Pitch
CSD	Constant Speed Drive
CSU	Constant Speed Unit
DEEC	Digital Electronic Engine Control
EBHA	Electric Back-up Hydraulic Actuator
ECIU	Engine-Cockpit Interface Unit
ECAM	Electronic Centralized Aircraft Monitor
ECS	Environmental Control System
EDP	Engine Driven Pump
EDU	Engine Display Unit
EEC	Electronic Engine Control
EFMPS	Electric Fuel Pumping & Metering System
EHA	Electro Hydrostatic Actuator
EHD	Elasto-Hydro-Dynamic
EHSV	Electro-Hydraulic Servo Valve
EICAS	Engine Indication and Caution Advisory System
EMI	Electro-Magnetic Interference
EPR	Engine Pressure Ratio

FAA	Federal Airworthiness Authority
FADEC	Full Authority Digital Electronic Control
FMU	Fuel Metering Unit
FRTT	Fuel Return To Tank
IEPR	Integrated Engine Pressure Ratio
HBV	Handling Bleed Valve
ICAO	International Civil Aviation Organization
IBV	Interstage Bleed Valve
IDG	Integrated Drive Generator
IGV	Inlet Guide Vanes
IP	Intermediate Pressure
HIRF	High Intensity Radiated Frequencies
HP	High Pressure
LP	Low Pressure
LVDT	Linear Variable Differential Transformer
MCL	Maximum Climb
MCR	Maximum CRuise
MEA	More Electric Aircraft
MEE	More Electric Engine
MR	Maximum Reverse
MTO	Maximum Take-Off
NGS	Nitrogen Generation System
NTSB	National Transportation Safety Board
OLTF	Open Loop Transfer Function
O&M	Overhaul & Maintenance
PCU	Propeller Control Unit
PEC	Propeller Electronic Control
PEM	Power Electronic Module
PHM	Prognostics and Health Monitoring
PLA	Power Lever Angle
PLF	Pressure Loss Factor
PMA	Permanent Magnet Alternator
PTIT	Power Turbine Inlet Temperature
R&O	Repair & Overhaul
RAT	Ram Air Turbine
RTD	Resistance Temperature Device
SD	Shut-Down
SFAR	Special Federal Airworthiness Regulation
SHP	Shaft Horsepower
SLS	Sea Level Static
SOV	Shut-Off Valve
STOVL	Short Take-Off and Vertical Landing
teos	Technology for Energy Optimized Aircraft Equipment & Systems
TGT	Turbine Gas Temperature
TIT	Turbine Inlet Temperature
TM	Torque Motor

TRU	Transformer Rectifier Unit
VIF	Vectoring In Flight
VLSI	Very Large Scale Integration
VSCF	Variable Speed Constant Frequency
VSTOL	Vertical or Short Take-Off and Landing
VSV	Variable Stator Vane
UAV	Unmanned Air Vehicle

1

Introduction

The modern gas turbine engine used for aircraft propulsion is a complex machine comprising many systems and subsystems that are required to operate together as a complex integrated entity. The complexity of the gas turbine propulsion engine has evolved over a period of more than 70 years. Today, these machines can be seen in a wide range of applications from small auxiliary power units (APUs) delivering shaft power to sophisticated vectored thrust engines in modern fighter aircraft.

The military imperative of air superiority was the driving force behind the development of the gas turbine for aircraft propulsion. It had to be lighter, smaller and, above all, it had to provide thrust in a form which would allow higher aircraft speed. Since aircraft propulsion is, by definition, a reaction to a flow of air or gas created by a prime mover, the idea of using a gas turbine to create a hot jet was first suggested by Sir Frank Whittle in 1929. He applied for and obtained a patent on the idea in 1930. He attracted commercial interests in the idea in 1935 and set up Power Jets Ltd. to develop a demonstrator engine which first ran in 1937. By 1939, the British Air Ministry became interested enough to support a flight demonstration. They contracted Power Jets Ltd. for the engine and the Gloucester Aircraft Co. to build an experimental aircraft. Its first flight took place on 15 May 1941. This historic event ushered in the jet age.

1.1 Gas Turbine Concepts

Operation of the gas turbine engine is illustrated by the basic concept shown schematically in Figure 1.1. This compressor-turbine 'bootstrap' arrangement becomes self-sustaining above a certain rotational speed. As additional fuel is added speed increases and excess 'gas horsepower' is generated. The gas horsepower delivered by a gas generator can be used in various engine design arrangements for the production of thrust or shaft power, as will be covered in the ensuing discussion.

In its simplest form, the high-energy gases exit through a jet pipe and nozzle as in a pure turbojet engine (the Whittle concept). This produces a very high velocity jet which, while compact, results in relatively low propulsion efficiency. Such an arrangement is suitable for high-speed military airplanes which need a small frontal area to minimize drag.

Gas Turbine Propulsion Systems, First Edition. Bernie MacIsaac and Roy Langton.
© 2011 John Wiley & Sons, Ltd. Published 2011 by John Wiley & Sons, Ltd.

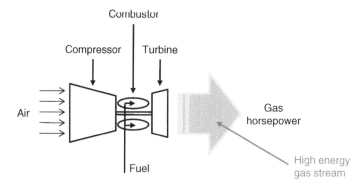

Figure 1.1   Gas turbine basics – the gas generator.

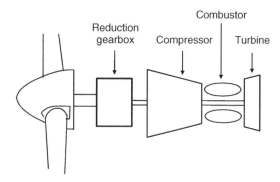

Figure 1.2   Typical single-shaft engine arrangement.

The next most obvious arrangement, especially as seen from a historical perspective, is the single-shaft turbine engine driving a propeller directly (see the schematic in Figure 1.2). As indicated by the figure the turbine converts all of the available energy into shaft power, some of which is consumed by the compressor; the remainder is used to drive the propeller. This arrangement requires a reduction gearbox in order to obtain optimum propeller speed. Furthermore, the desirability of a traction propeller favors the arrangement whereby the gearbox is attached to the engine in front of the compressor.

The Rolls-Royce Dart is an early and very successful example of this configuration. This engine comprises a two-stage centrifugal compressor with a modest pressure ratio of about 6:1 and a two-stage turbine. The propeller drive is through the front of the engine via an in-line epicyclic reduction gearbox. The Dart entered service in 1953 delivering 1800 shaft horsepower (SHP). Later versions of the engine were capable of up to 3000 SHP and the engine remained in production until 1986.

Today, single-shaft gas turbines are mostly confined to low power (less than 1000 SHP) propulsion engines and APUs where simplicity and low cost are major design drivers. There are some notable exceptions, however, one of which is the Garrett (previously Allied Signal and now Honeywell) TPE331 Turboprop which has been up-rated to more

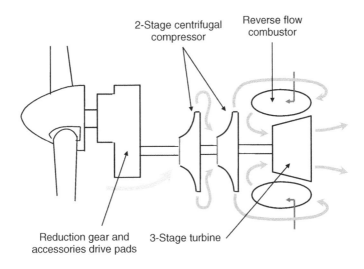

Figure 1.3   TPE331 turboprop schematic.

than 1600 SHP and continues to win important new programs, particularly in the growing unmanned air vehicle (UAV) market.

This engine is similar in concept to the Dart engine mentioned above, as illustrated by the schematic of Figure 1.3. The significant differences are the reverse-flow combustor which reduces the length of the engine and the reduction gear configuration which uses a spur-gear and lay-shaft arrangement that moves the propeller centerline above that of the turbine machinery, thus supporting a low air inlet.

A more common alternative to the direct-drive or single-shaft arrangement described above uses a separate power turbine to absorb the available gas horsepower from the gas generator.

Since the power turbine is now mechanically decoupled from the gas generator shaft, it is often referred to as a 'free turbine'.

For the purposes of driving a propeller, this configuration (as shown in Figure 1.4) indicates a requirement for a long slender shaft driving through a hollow gas turbine shaft to the front-mounted gearbox. Such a configuration carries with it the problems of shaft stability, both lateral and torsional, together with more complex bearing arrangements.

In their turboprop concept, Pratt & Whitney Canada chose to 'fly the engine backwards' by arranging for sophisticated ducting for the inlet and exhaust while benefitting from the stiffness and robustness of a very short drive shaft through a reduction gearbox. Their engine, the PT-6 in its many configurations, is one of the most reliable aircraft gas turbines ever built. It has an exceedingly low in-flight incident rate and has sold over 40 000 copies. It was first introduced in 1964 and is still very much in production. A conceptual drawing of the PT-6 engine is shown in Figure 1.5.

The pure turbojet produces a high-velocity jet, which offers poor propulsion efficiency with the singular advantage of higher aircraft speed, and the turboprop produces good propulsive efficiency but only at a relatively low top aircraft speed. The two configurations can however be combined to produce the turbofan engine, depicted in Figure 1.6. As is

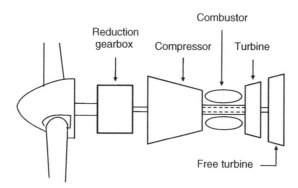

Figure 1.4   The free turbine turboprop engine.

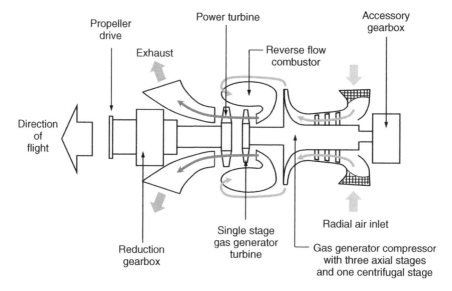

Figure 1.5   A sectional drawing of the PWA PT-6 turboprop engine.

indicated in this figure, the front-mounted fan is driven by a shaft connected through the core of the engine to the second or low-pressure turbine which can be likened to the free turbine of the turboprop application. Some of the fan flow pressurizes the compressor while the remainder is expelled through a so-called 'cold nozzle' delivering thrust directly. Such an arrangement can produce high thrust and good propulsive efficiency, and this engine concept is one of the most common types in commercial service today.

   Another important configuration used in aircraft propulsion is the twin-spool turbojet engine which is essentially a twin-spool gas generator with a jet pipe and exhaust nozzle. If a second turbine can drive a large fan, it can also drive a multistage compressor with an output which is entirely swallowed by the downstream compressor. This configuration is shown in Figure 1.7.

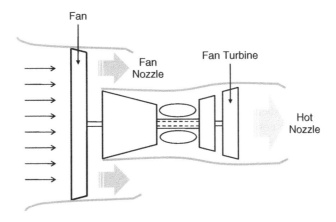

Figure 1.6   The turbofan engine configuration.

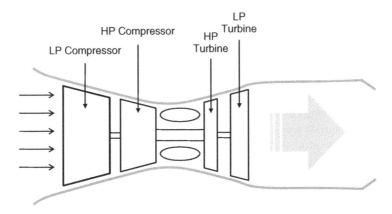

Figure 1.7   The twin-spool turbojet engine configuration.

So far in this discussion, we have assumed that the thermodynamic processes of compression and expansion are ideal and that there is no apparent limit to the magnitude of
the pressure that can be obtained. In addition, we have not considered how the heat is
going to be delivered to the gas to raise its temperature.

The practical implementation of the gas turbine involves turbomachinery of finite
efficiency and an internal combustion process that adds heat through the burning of a
hydrocarbon fuel in a combustion chamber which must be small and compact.

Throughout its development, there have been enduring themes which place specific
technologies in the vanguard of engine development. The first of these themes is engine
performance: the capacity of the engine to produce thrust with sufficient thermal efficiency to provide an airplane with an acceptable range while carrying a useful payload.
The response to this requirement is found in the techniques of internal aerodynamics
and combustion.

Saravanamuttoo *et al.* [1] provide a comprehensive treatment of gas turbine performance. Simple cycle calculations highlight the need for high overall engine pressure ratios and high turbine temperatures for good efficiency to be achieved. Similarly, high specific thrust demands high isentropic efficiency of each major component. Finally, size matters. In order to achieve high levels of thrust, high air flow rates must be obtained. This argues powerfully for large axial flow turbomachinery. This is very much a pacing item, since the design of such machines is very complex and the investments in equipment and facilities required to complete the development are very large indeed.

A similar argument can be made for combustion technologies. The compressor must deliver a uniform flow of air at high pressure to a combustion chamber. Fuel must be introduced into the combustor in sufficient quantities to raise the average temperature by at least 1200 °F. Assuming that the combustion process takes place at nearly stoichiometric conditions, localized temperatures in excess of 3500 °F can be expected. Excess air is essential in the gas turbine combustor to cool the flame to acceptable levels while, at the same time, mixing the hot gas to deliver a uniform, high-temperature gas to the throat of the turbine. Finally, in the interests of weight and overall engine stiffness and robustness, the combustor must be kept as short as possible. Again, this is a technology which relies heavily on experiment which, in turn, involves large investments in equipment and facilities.

The second major theme that runs throughout the development of the jet engine is that of longer life and improved reliability. This requirement has driven a relentless quest for improved materials and design methodologies. The basic need is for turbine components capable of operating continuously at elevated temperatures. (Turbine inlet temperatures for uncooled blades can run as high as 2500 °F.) Both blades and disks must be capable of withstanding the enormous stresses imposed by rotational speeds which push the materials past the elastic limit, thereby encountering low cyclic fatigue. This must be understood well enough to ensure reasonable life as well as removal before safety concerns overtake them.

The twin themes of continuous improvements in aerothermodynamics and in materials would suggest that the gas turbine engine, while sophisticated, is actually a very simple machine. In fact, the quest for improved performance has led designers to a remarkable number of variations in engine configuration. Each configuration, when matched to the airframe for which it was designed, offers a different balance between fuel efficiency, specific thrust and overall propulsive efficiency. Single-, twin- and triple-spool engine configurations have been developed with attendant increases in the complexity of bearing and lubrication systems. The turbofan engine has become the workhorse of the civil aviation industry with sophisticated thrust management, including thrust reversal and power extraction to drive a variety of accessories. The gas turbine engine has therefore emerged as a sophisticated and complex machine requiring a systems approach to its design and development.

1.2 Gas Turbine Systems Overview

In order to provide the reader with a basic knowledge, the gas turbine engine aerothermodynamic principles described in Chapter 2 of this book provide insight into some of the challenges associated with the fundamentals of gas turbine design, operation and control. A more detailed treatment of axial compressor design concepts, including compressor performance analysis and the principles of compressor performance map estimation, are included as Appendices A and B, respectively. For completeness, thermodynamic modeling of the gas turbine engine is described in Appendix C.

While there are many systems and subsystems that make up the gas turbine-based propulsion power plant, perhaps the most critical function is performed by the fuel control system.

This system must provide high-pressure fuel to the combustor of the gas generator or 'core' section of the engine over the complete operational envelope, while protecting the machine from temperature, pressure and speed exceedances for any combination of dynamic and steady-state operation.

In addition, the fuel control system may be required to manage airflow though the compressor by modulating compressor stator vanes and bleed valves.

The gas generator produces high-energy gases as its output, sometimes referred to as gas horsepower or gas torque, which can be converted into direct thrust or shaft power.

In military aircraft with thrust augmentation (afterburning), the fuel control system is also required to control afterburner fuel delivery together with the control of exhaust nozzle exit area in order to maintain stable gas generator operation.

Secondary functions of the fuel control system include cooling of the engine lubricating oil and, in some applications, providing a source of high-pressure fuel to the airframe to act as motive flow to the aircraft fuel system ejector pumps [2].

In view of the complexity and extent of the fuel control system issues, this important topic is covered in three separate chapters as follows.

1. The fuel control of the gas generator section, including acceleration and deceleration limiting, speed governing and exceedance protection, is covered in Chapter 3.
2. Thrust engine fuel control issues, including thrust management and augmentation, are described in Chapter 4.
3. Fuel control and management of shaft power engines, including turboprop and turboshaft applications, are presented in Chapter 5.

Since major performance issues associated with fuel control systems design involve dynamic response and stability analyses, Appendix D is provided as a primer on classical feedback control.

In commercial aircraft it is standard practice to install many of the engine subsystems and associated major components as part of an engine, nacelle and strut assembly. This integrated nacelle/engine package is then delivered to the airframe final assembly line for installation into the aircraft.

For reasons of aerodynamic performance or stealth, military aircraft are more likely to integrate the propulsion system assembly more closely with the fuselage.

While the primary function of the engine installation arrangement is to provide efficient and effective air inlet and exhaust for the gas turbine engine, provisions for minimizing engine compressor noise propagation as well as ventilation and cooling of the installation must also be considered. The thrust reversing mechanism, including actuators and nozzle flow diversion devices, is also typically installed at the nacelle or propulsion system assembly stage.

Supersonic applications present a special case to the propulsion system designer. Here the task of recovering free stream energy efficiently to the engine inlet face requires the management of shock-wave position within the inlet through the control of inlet geometry. While supersonic inlet control is often included as an airframe responsibility, it is nevertheless a major factor is providing efficient propulsion in supersonic flight and is therefore addressed in this book.

Installation-related systems issues, focusing primarily on inlet and exhaust systems, are presented in Chapter 6.

As with any high-power rotating machine, bearing lubrication and cooling is a critical function and the task is further complicated by the operational environment provided by an aircraft in flight. Chapter 7 addresses the primary issues associated with lubrication systems of aircraft propulsion gas turbines engines.

In addition to providing propulsion power in aircraft applications, the gas turbine engine must also provide a source of power for all of the energy-consuming systems on the aircraft. This power is removed from the engine in two forms, as described below.

- Mechanical power is taken from the shaft connecting the turbine and compressor. This power source, which involves a tower shaft and reduction gearbox, shares the engine lubrication system. A number of drive pads are typically provided for electrical generators and hydraulic pumps. Engine starting is effected through this same gearbox
- Bleed air is also used by the airframe for cockpit/cabin pressurization and air conditioning. This source of hot high-pressure air is also used for anti-icing of both the wing and engine nacelle air inlet.

The systems, subsystems and major components associated with mechanical and bleed air power extraction and starting systems are covered in Chapter 8.

So far we have considered gas turbines in aircraft applications only. In the defense industry, however, the benefits of the gas turbine in terms of power per unit weight have not gone unnoticed. Many of today's high-speed naval surface vessels use the gas turbine as the main propulsion device. For completeness, marine gas turbine propulsion systems focusing on naval applications are therefore included in Chapter 9.

The issue of prognostics and health monitoring (PHM) has become a critical issue associated with in-service logistics over the past several years; both the commercial airlines and military maintenance organizations are moving away from scheduled maintenance to on-condition maintenance as a major opportunity to improve efficiency and reduce the cost of ownership.

Chapter 10 describes PHM, covering the basic concepts of engine maintenance and overhaul strategies and the economic benefits resulting from their application. Also addressed are the techniques used in the measurement, management and optimization of repair and overhaul (R&O) practices for application at the fleet level.

Finally, some of the new system technologies that are being considered for future gas turbine propulsion systems are discussed in Chapter 11. Of particular interest by many engine technology specialists is the 'more-electric engine' (MEE) initiative which is an offshoot from what began as the 'all-electric aircraft' (now the 'more-electric aircraft') launched by the Wright Patterson Air Force Laboratory some 40 years ago.

References

1. Saravanamuttoo, H.I.H., Rogers, G.F.C, and Cohen, H. (1951–2001) *Gas Turbine Theory*, 5th edn, Pearson Education Ltd.
2. Langton, R., Clark, C., Hewitt, M., and Richards, L. (2009) *Aircraft Fuel Systems*, John Wiley & Sons, Ltd, UK.

2

Basic Gas Turbine Operation

The focus of this book is the many systems that are needed in order to produce a successful propulsion engine. However, no treatment of a system design can proceed without an examination of the fundamentals of engine behavior which informs the designer of the basic requirements of the machine for which the system is intended.

This chapter is therefore devoted to an examination of the operating characteristics of the gas turbine. The focus will be on the practical features of the major components and the impact that these features have on the operation of the engine as a system. By inference, they offer some insight into why so many configurations of the gas turbine have been developed.

In purely thermodynamic terms, every gas turbine is a practical implementation of the classic Brayton cycle. This cycle is commonly presented on a temperature–entropy (T–S) diagram as shown in Figure 2.1.

The cycle begins at condition 1, at which point the gas goes through a pure compression phase to a higher temperature and pressure at state 2. The work done by the compressor is represented by the change of thermodynamic state defined by higher pressure and temperature.

Heat is then added to the gas at constant pressure raising its temperature to that of state 3. From state 3 to state 4, the gas is expanded back to a pressure defined by state 1. If the expansion is done through a turbine, sufficient power can be extracted to drive the compressor with enough left over to drive another device such as a propeller. Alternatively, the energy left over from driving the compressor can be expanded through a propelling nozzle to produce thrust.

2.1 Turbojet Engine Performance

The single-spool turbojet is generally regarded as the simplest form of gas turbine. The possibilities for the inclusion of variable geometry will be neglected for the moment and the engine will be considered to have a single moving part: a rotor comprising a compressor and turbine. Its operation is described as follows.

- Air enters the engine at the face of the compressor at conditions of pressure and temperature defined by P_2 and T_2 (atmospheric state) and is compressed to a new thermodynamic state defined by pressures and temperatures P_3 and T_3.

Gas Turbine Propulsion Systems, First Edition. Bernie MacIsaac and Roy Langton.
© 2011 John Wiley & Sons, Ltd. Published 2011 by John Wiley & Sons, Ltd.

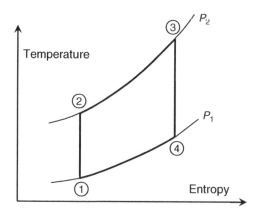

Figure 2.1 The ideal Brayton cycle.

- Fuel is mixed with the compressor air and burned, thereby raising the temperature of the mixture to the highest temperature T_4 permitted by available materials.
- The hot gas is then expanded through the turbine, producing sufficient power to drive the compressor.
- The gas exiting the turbine (still very energetic) is further expanded through the propelling nozzle which produces the thrust needed to drive the airplane.

A diagram of this operation is provided in Figure 2.2. This figure indicates the stations along the gas path of the engine, at which points thermodynamic states are defined. The stage numbering scheme shown is common to many gas turbine engines; however, it is simply a convention and can change with different engine design concepts.

The previous description can be found in just about any treatment of the gas turbine engine. Publications [1] and [2] are examples produced by engine manufacturers Pratt & Whitney and Rolls-Royce, respectively. A third book by Irwin E. Treager [3] is also recommended as an abundant source of gas turbine engine information.

It must be recognized, however, that our interests here are focused on the systems aspect of the engine. We therefore want to explore the interaction of these components with each other and the engine with its environment. Let us therefore begin with an analysis of the performance of a turbojet engine both at design and then off-design conditions.

We begin this topic by first recognizing that compressors and turbines are aerodynamic components which are individually the subject of a design effort. Following the selection of the overall engine design parameters of pressure ratio and turbine inlet temperature, the design point of each of the major components is specified by cycle calculations. Leaving aside issues of allowable physical envelope for the moment, a compressor will be defined by the following aerodynamic parameters:

- air flow rate;
- pressure ratio; and
- isentropic efficiency.

A similar set of parameters will define the design point of the turbine.

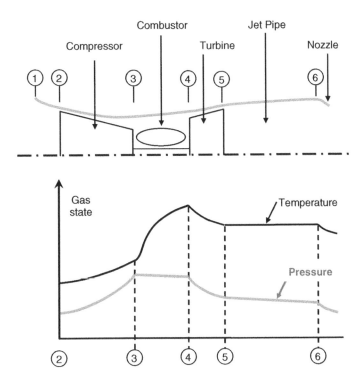

Figure 2.2   Typical turbojet application.

Due to the magnitude and complexity of the design task associated with a multistage axial compressor, the work is typically broken down into specialist activities. For example, a compressor aerodynamics group will undertake to specify the number of compressor stages and the size and profile of blade shapes for each stage. From this analysis, a number of parameters will be selected including rotational speed, annulus size, blade stagger, and so on. These parameters, in turn, drive the mechanical design which will produce a physical embodiment which must be lightweight and reliable. A similar effort by the turbine design group will produce a similar design for the turbine.

The reader is reminded that the jet engine is a prime mover which must start from zero speed, accelerate to idle conditions and operate at any point between idle and full power at any altitude and, in some military applications, in any attitude. Each component design group must therefore explore, using a combination of analysis and test, the off-design behavior of the component in question. For compressors and turbines, the results are in the form of performance 'maps' which describe the full envelope of behavior for that component. Typical compressor and turbine maps are shown in Figure 2.3.

It should be noted that these maps are presented in non-dimensional form. The non-dimensional parameters are derived from the Buckingham Pi theorem, which recognizes that any complex system involving a number of variables can be represented by a set of non-dimensional parameters that is always less than the number of variables by the number of dimensions used. Thus, the variations in compressor and turbine performance

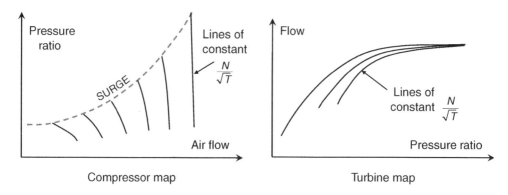

Figure 2.3 Typical compressor and turbine flow performance maps.

which are associated with changes in altitude are conveniently represented by normalizing the maps to sea-level conditions in non-dimensional form.

The full non-dimensional form involves a characteristic dimension such as the inlet diameter; however, this is constant for a given machine so it is commonly dropped from the term. The performance parameters are therefore as follows:

$$\text{Airflow:} \quad \frac{w\sqrt{T}}{D^2 P} \Rightarrow \frac{w\sqrt{T}}{P}$$

$$\text{Speed:} \quad \frac{ND}{\sqrt{T}} \Rightarrow \frac{N}{\sqrt{T}}$$

$$\text{Pressure ratio:} \quad \frac{P_3}{P_2} \Rightarrow \frac{P_3}{P_2}$$

$$\text{Efficiency:} \quad \frac{T_3' - T_2}{T_3 - T_2} \Rightarrow \eta$$

where T_3' is the ideal temperature at P_3, w is the engine airflow rate, D is a diameter, (usually taken as the diameter of the front of the compressor), and N is the rotational speed of the engine rotor.

In this form, the effect of air density with altitude is accommodated as is the effect of forward speed by referencing inlet conditions (pressure, temperature) to the face of the engine.

The assembly of these major components into a jet engine as shown previously in Figure 2.2 imposes two specific constraints on the operation of the compressor and turbine.

1. All of the air from the compressor must pass through the combustor and turbine. For steady-state conditions, this is commonly referred to as 'compatibility of flow'.
2. The power generated by the turbine is absorbed by the compressor. In steady state, this is referred to as 'compatibility of work'.

For a given propelling nozzle area, the above constraints force the turbine and compressor to match each other such that there is a single overall pressure ratio and turbine

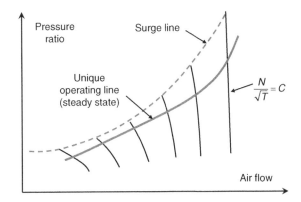

Figure 2.4 Steady-state operation of the compressor of a turbojet engine.

inlet temperature at which the engine will operate in steady state for a given rotor speed. The idea of a unique operating point can therefore be employed to describe the operation of the engine. A locus of such points can be drawn on the compressor map as shown in Figure 2.4, and is commonly referred to as the 'engine operating line' or 'steady running line'. The unique operating line concept breaks down if a variable area nozzle is employed, of course. This will be discussed more fully in Section 2.1.3.

Since fuel control is one of the systems to be covered in this book, this is an appropriate point to address the dynamic behavior of the engine. Let us therefore consider the consequences of a change in the fuel flow on engine operation. At the outset, it would be fair to say that a positive change in fuel flow rate will tend to increase the power level. It will also increase the rotational speed of the engine which, in turn, means the air flow rate will change. It therefore follows that the steady-state conditions of compatibility flow and work will be upset. In other words, the engine will be operating in a dynamic state.

Let us consider the operation of the combustion chamber, which is shown diagrammatically in Figure 2.5.

In this figure, the combustor should be thought of as an accumulator whose pressure level is dictated by the law of conservation of mass. The rate of change of density within the combustion is therefore given by:

$$\frac{d\rho_3}{dt} = \frac{1}{V}(w_3 - w_4 + w_{\text{Fe}})$$
(2.1)

where V is the volume of the combustor can and w_{Fe} is the fuel flow rate.

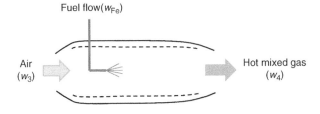

Figure 2.5 Combustion chamber operation.

Now, the gas law can be written in differential form as follows:

$$\frac{dP}{dt} = RT\frac{d\rho}{dt} + \rho R\frac{dT}{dt}. \tag{2.2}$$

For the present analysis, we will neglect the second term on the right because its effect is very much weaker than the first term and good results can be obtained with this simplification. We can therefore write the conservation of mass equation as follows:

$$\frac{dP_3}{dt} = \frac{RT}{V}(w_3 - w_4 + w_{\text{Fe}}). \tag{2.3}$$

A casual examination of this equation would suggest a 25% increase in fuel flow would influence the rate of change of pressure by less than 1% since fuel/air ratio is of the order 0.03. The real influence, however, is obtained from the fact that, as the fuel is consumed, there is an attendant increase in the temperature rise across the combustor associated with the increased fuel burned. The balance of energy in the combustor can be expressed as follows:

$$h_3 w_3 + \Delta H w_{\text{Fe}} = h_4 w_4. \tag{2.4}$$

Again, in simplifying this equation by setting $w_4 = w_3$ we obtain:

$$h_4 - h_3 = \Delta H \frac{w_{\text{Fe}}}{w_3} \tag{2.5}$$

or, in terms of temperature, we can write:

$$T_4 - T_3 = \frac{\Delta H}{c_{\text{p}}} \frac{w_{\text{Fe}}}{w_3} \tag{2.6}$$

where c_{p} is the specific heat of air. For most practical purposes, the chemical process of heat release is very nearly instantaneous. This rapid change in temperature T_4 thus affects the turbine flow w_4 because the parameter $w_4\sqrt{T_4}/P_4$ controls the flow through the turbine nozzle area.

Referring to Equation 2.3, which describes the rate of change of combustor pressure, it would appear that the dominant influence on P_3 is the reduction in combustor exit flow rate. In fact, applying representative values for the compressor inlet temperature and the physical volume of the combustor, we can estimate that the rate of change in pressure is of the order 1000 psi/s. This translates to a first-order time constant for Equation 2.3 of the order a few milliseconds.

Simultaneously with the upset in the compatibility of flow, the power on the shaft is unbalanced in favor of the turbine. The rate of change of rotor speed can therefore be estimated as follows:

$$\frac{dN}{dt} = \frac{2\pi}{I_{\text{g}}}(G_{\text{t}} - G_{\text{c}}) \tag{2.7}$$

where I_{g} is the polar moment of inertia of the rotor, G_{t} is turbine torque and G_{c} is compressor torque.

Applying typical values for these parameters suggests that the rate of change of rotor speed to torque imbalance is very much slower than the rate of change of combustor pressure. Again, expressing the response rate in terms of the first-order time constant of Equation 2.7, we obtain values of the order 0.5–1 s for sea-level operating conditions.

If we were to plot the operating point of the compressor through this transient state, we could show that the pressure ratio will rise very rapidly compared to the rate of change of rotor speed. In effect, the compressor operating point will migrate from the steady-state operating line along a steady-state speed line in the direction of surge as shown in Figure 2.6.

The fact that the response rate of compressor delivery pressure to a step increase in fuel flow rate is about a thousand times faster than the response rate of engine rotor speed suggests that the operating point of the compressor will simply migrate into the surge region of the performance map as shown in Figure 2.6, unless the fuel increase is limited in some way.

The phenomenon of compressor surge or stall is associated with a very rapid collapse of flow rate through the machine. This results in violent changes in forces on individual blades typically at quite high frequencies which, if continued, will cause mechanical destruction of the compressor. The photograph in Figure 2.7 is an example of the consequences of an uncontrolled engine surge.

We therefore encounter one of the most intractable control problems faced by engineers responsible for the development of the jet engine: compressor stall. Because the root cause of compressor stall is bound up in the turbulent boundary layer and its interaction with the shape and condition of individual compressor blades, there is a stochastic element to the phenomenon. Machining tolerances matter as does the general condition of the blade insofar as erosion, corrosion, and cleanliness is concerned. We refer to the difference between the maximum compressor delivery pressure that can be obtained at a given rotor speed and the steady-state operating point at the same speed as the 'available surge margin'. It is the available surge margin of an engine that allows us to increase the fuel flow rate which, in turn, causes the engine to accelerate to a higher power level. One of

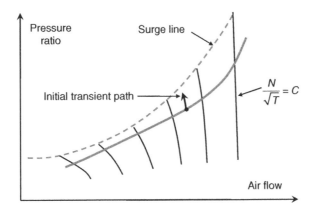

Figure 2.6   Compressor operating point transient following a step increase in fuel flow.

Figure 2.7 Photograph of compressor damage following a surge (courtesy of Ian Nunn).

the principal functions of the fuel control is to manage the fuel delivered to the engine throughout the transient, such that compressor stall is avoided.

2.1.1 Engine Performance Characteristics

Once the major components of an engine have been 'matched' to each other and their respective operating points are known over their full envelope, it becomes possible to characterize the overall performance of the engine. Again, these are presented in non-dimensional terms.

Perhaps the most important parameters from a system management perspective are fuel flow, rotational speed, and a parameter which describes overall thrust or is a surrogate for thrust. These parameters are described as follows.

2.1.1.1 Fuel Flow Rate

The full non-dimensional form for fuel flow rate is:

$$\frac{w_{Fe}\Delta H}{D^2 P \sqrt{T}}$$

where w_{Fe} is the mass fuel flow rate, ΔH is the calorific value of the fuel, D is the nominal diameter of the engine, T is the temperature at the engine inlet, and P is the pressure at the engine inlet.

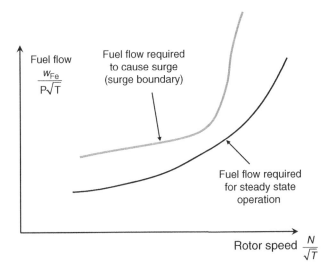

Figure 2.8   Engine fuel versus speed performance.

Since the calorific value of the fuel is normally considered a constant for a given fuel type and the nominal diameter of the engine is fixed, the fuel flow parameter is usually expressed as:

$$\frac{w_{Fe}}{P\sqrt{T}}.$$

Similarly, the rotor speed is expressed non-dimensionally as $N/\sqrt{T}$.

It is therefore possible and conventional to plot fuel flow versus rotor speed in non-dimensional form as shown in Figure 2.8.

As shown in the figure, the steady-state curve defines the quantity of fuel required to maintain engine operation at a specified rotational speed. The other curve is, in some respects, more interesting. This curve is a quasi-steady-state description of the amount of fuel (per unit time) in non-dimensional form required to drive the engine into a state of compressor stall from a given steady-state speed condition.

This curve can be determined experimentally by allowing the engine to stabilize at a specific rotor speed and then to impose a sudden increase in fuel without the benefit of control protection. If the sudden increase in fuel is just large enough, the compressor will stall and a single point on the stall fuel boundary curve will have been established. By exploring a number of steady-state speed points in this manner, the entire stall fuel boundary can be mapped.

An alternative method of testing involves a variable geometry propelling nozzle. A single-spool engine responds to changes in exhaust nozzle area by migration of the compressor operating point up or down a speed line, depending on the direction of the change [1].

Specifically, a smaller nozzle will cause the operating point to move toward surge. Thus, the engine can be stabilized at a given rotational speed and the nozzle area reduced. As the nozzle area is reduced, the operating point will move toward surge and the rotational speed will change for fixed fuel flow rate. As the compressor approaches stall, the fuel flow rate and the rotational speed can be noted and recorded. Obviously, it is possible to hold the speed constant by appropriate adjustment of the fuel flow rate at each new nozzle setting, thereby obtaining a point stall fuel boundary at fixed rotor speed.

Since such engine tests are both expensive and dangerous, these data points can be estimated using a computer model of the engine which is detailed enough to contain descriptions of the component maps. It is therefore a simple calculation to fix the engine speed and increase fuel flow to cause the operating point to migrate toward compressor stall. Such a model is described in Appendix C.

Since the quantity of fuel defined by the stall fuel boundary curve is fixed by the tendency for the compressor to stall, it follows that the shape of the stall fuel boundary curve is inextricably linked to the shape of the compressor surge or stall boundary which, in turn, is related to its design.

The design of an aircraft gas turbine compressor that has a wide speed margin allowing stable operation at idle conditions is, at best, a struggle. Many compressors offer very little stall margin at low speeds due to the interaction of compressor stages which tends to drive the front stages into stall. Designers of such machines often resort to measures to relieve the front stages at low speed. This can be achieved by resetting the stagger of the stator blades or by bleeding away some air part-way along the compressor, which has the effect of allowing the front stages to swallow more air thus relieving their propensity to stall.

In either case, these control measures suggest a likely reduction in stall margin as the bleed valve is closed or the stators move to their high-power operating condition. In the case where the control action is abrupt, this can have a very sharply defined 'dog-leg' in the stall fuel boundary. In other cases, notably on engines designed by General Electric, the control is more continuous and smooth resulting in more tractable surge margins from the perspective of the control's designer.

2.1.1.2 Engine Thrust

The non-dimensional form for thrust produced by a jet engine is

$$\frac{F}{D^2 P}$$

where F is gross thrust, D is characteristic dimension (typically engine inlet diameter) and P is engine inlet pressure.

A typical plot of this parameter versus non-dimensional rotor speed is depicted in Figure 2.9.

This parameter is readily measured at sea-level static conditions in a properly calibrated test cell; however, its measurement in-flight is not a simple matter. Engine parameters other than gross thrust are therefore used for setting throttles on takeoff and landing and for thrust management during flight operations.

Firstly, of primary importance from a systems perspective, propulsive thrust is a function of the *net* change in momentum of the air entering the engine and is therefore dependent on aircraft speed. This fact complicates any in-flight measurement of net thrust.

Secondly, there are compelling safety reasons why the pilots of an aircraft must keep the thrust of the engine within specified limits. These are dominated by the thrust available at takeoff and the thrust available during final approach and landing.

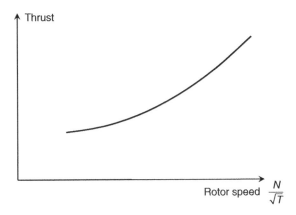

Figure 2.9   Engine gross thrust versus speed.

Finally, there are operational reasons for matching engine thrust to aircraft require-ments as accurately as possible throughout the flight envelope. For example as altitude is increased, the density of the air decreases, demanding less thrust for given aircraft speed. Similarly as fuel is burned off during the flight, aircraft weight decreases which again reduces the thrust required for a given speed. High thrust settings, which require opera-tion at engine temperatures beyond what is immediately needed, will ultimately reduce the useful life of the engine; thus, power management systems provide for continuous adjustments of the power level to match flight conditions.

For purposes of control, there are three parameters commonly in use. These are:

- engine pressure ratio (EPR);
- integrated engine pressure ratio (IEPR); and
- corrected fan speed ($N/\sqrt{T}$).

The first of these (EPR) is defined as the ratio of turbine exhaust pressure to engine inlet pressure. This parameter approximates the pressure ratio across the nozzle, especially at the start of the takeoff run (i.e., zero forward speed) and is therefore related to the gross thrust of the engine. Both exhaust pipe pressure and engine inlet pressure are readily measured, making this parameter a convenient, albeit rough, measure of thrust. As forward speed increases, this parameter requires correction in order to obtain a good correlation with actual thrust. This was not a difficulty for operators in the early days and is readily handled by modern power management controls today.

The second parameter (IEPR) is related to the first, but has been adopted by one manufacturer to account for the fact that modern engines are turbofans and that a large fraction of the total thrust is obtained from the fan airflow which does not enter the core engine. There are therefore two jet pipes and two nozzles which, in general, will be oper-ating at different pressure ratios. The IEPR is an algebraic expression which attempts to approximate a single overall average pressure ratio which, in turn, will provide a measure of engine thrust. This parameter involves the measurement of three pressures: engine inlet, fan outlet or cold nozzle jet pipe, and turbine exit pressure or hot nozzle jet pipe.

Finally, another manufacturer has also observed that most of the thrust from a modern high bypass ratio turbofan engine is obtained from the fan airflow. This manufacturer has therefore chosen corrected fan speed as the parameter with which to set the thrust at takeoff and to manage the engine thrust in-flight.

We shall leave it to others to judge the efficacy of each of these parameters. None of them are a direct measure of thrust, but all of them work adequately well as a means of control. The fan speed parameter is also used, in conjunction with the core engine speed, to recognize erosion and related damage to the leading edge of fan blades; it is therefore a useful measure by which to gauge overall engine performance deterioration as the machine ages. This observation is derived from the fact that the two spools of the twin-spool engine (turbojet or turbofan) are coupled aerodynamically so that, for a given core engine speed, there is a unique fan speed at which the engine must operate. The aerodynamic coupling is directly related to the adiabatic efficiency and pumping capacity of each of the major components. If the thermodynamic behavior of one of these components were to change, the uniqueness of the rotor speed relationship also changes.

2.1.2   Compressor Surge Control

As described previously, modern efficient engines demand high pressure ratios and high turbine inlet temperatures [1]. This observation, coupled with high thrust requirements, leads us toward multistage axial flow compressors. While much progress has been made in the development of good stage efficiencies, the high overall pressure ratios required dictate a multistage design. Furthermore, since compressor stall is such a serious concern, especially for the controls designer, it is worthwhile developing some understanding of this phenomenon and the various means employed to improve surge margins during operation.

A multistage compressor is little more than a series of compressors stages stacked together on a common shaft. Each stage is therefore a compressor generating a pressure rise and attendant temperature rise. Each stage can be described by its own compressor map, although stage characteristics are normally presented in a somewhat different format. Finally, and of critical importance, the flow from each stage must be swallowed by the next stage so that each successive stage is sized to accommodate this flow and the flow should be presented to the next stage in a manner that optimizes stage efficiency. An example of a multistage compressor is shown in Figure 2.10.

Each compressor stage consists of a rotating set of blades mounted on the rim of the disc (see Figure 2.11). This is followed by a second set of stationary blades (commonly referred to as stators) whose task is to present the flow to the next stage in the best possible state.

By taking a 'top–down' view of the compressor blade set, it is possible to construct velocity diagrams associated with different points through the stage as shown in Figure 2.12.

In the case shown, the air enters the stage in a purely axial direction at velocity V_1. Because the first blade is attached to the rotating disc, it is moving with a tangential velocity of U. Using vector algebra and summing the absolute velocity V_1 with the blade velocity U, we obtain an air velocity relative to the blade: V_{r1}.

Using the conventional nomenclature for turbomachinery blading, the velocity vector V_{r1} can be seen to be at an angle of incidence i to the blade chord at the leading edge of

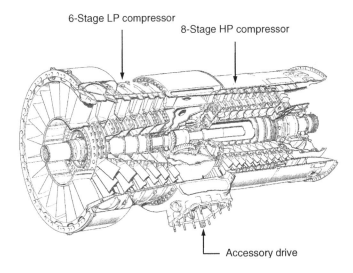

Figure 2.10 A typical multistage compressor (© Rolls-Royce plc 2011).

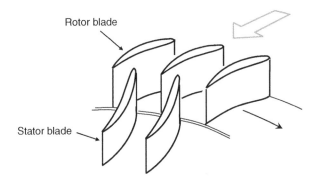

Figure 2.11 Typical compressor stage blade arrangement.

the blade. The rotating blade imparts considerable energy to the air and, in relative terms, it travels along the curved surface of the blade and exits at a relative velocity V_{r2} which is tangential to the blade chord at its trailing edge. From vector algebra, it is apparent that the absolute velocity V_2 will be as shown in the figure.

Finally, by orienting the blade of the stator so that its angle of incidence to the flow is within bounds, we can obtain a change in flow direction which then results in an acceptable angle of incidence for the next rotating blade. In the example shown, the air enters the first stage axially and the stator arranges to redirect the flow so that it also enters the second stage in an axial direction.

Common sense dictates that if the angle of incidence is too great (or too little in a negative sense), the blade will stall. Similarly, common sense also dictates that the area required to pass the flow from stage to stage must decrease in order to accommodate the higher level of compression delivered to the air by each stage.

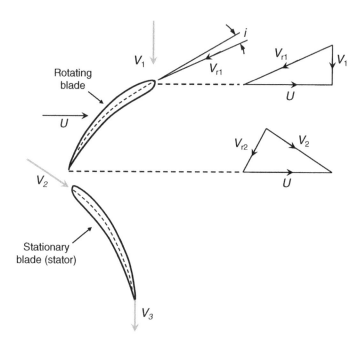

Figure 2.12   Axial compressor flow velocities.

These common sense observations indicate that the designer has to very carefully select the setting angle (stagger) of each row of blades and must select an annulus area at each stage to accommodate the flow. Overall, this is a multidimensional problem which is solved by a combination of sophisticated analysis and experimentation to arrive at a design which gives optimum efficiency and the desired pressure level with a minimum number of stages.

At the design point, the compressor will be operating at the best flow conditions possible for each stage. Let us now consider what happens as the design point is departed. An excellent treatment of this subject is provided by Stone [4] whose paper, although dated, is still considered one of the best descriptions available in the literature of the interaction of axial compressors stages. Stone's analysis shows clearly that, as a compressor operates at speeds between minimum and maximum, the operating point of each stage changes; at points below a specific intermediate speed, the front stages of the compressor are first to stall and above that speed, the rear stages are first to stall. This phenomenon is depicted in Figure 2.13.

At the higher speeds, all stages are highly loaded such that when the last stage stalls it always triggers overall compressor surge. This is shown on Figure 2.13 as point A. Just at or below the speed at which the front stage is first to stall, shown as point B in the diagram, it is highly likely that the stalling of the first stage will also dictate overall compressor stall. However, as the speed is reduced further, the front stage may stall but overall compressor surge may not occur. While not desirable, it is therefore possible that, at low speeds, there may be several of the front stages fully stalled but the compressor is still able to operate (albeit at a reduced efficiency). Point C on Figure 2.13 is the lowest

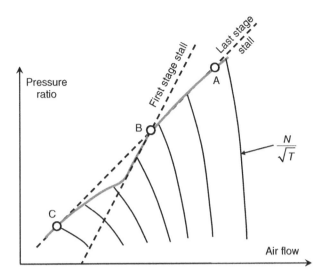

Figure 2.13   Compressor map showing stage stall and overall compressor surge.

speed and highest pressure ratios that can be sustained. This speed actually dictates the lowest possible idle speed for this compressor when operated as part of an engine.

The reader may observe at this point that if enough stages of an axial flow compressor stall, the machine may be prone to low-speed surge; worse, it may not be able to start because so many stages are simply unable to pass the required flow to support compression in the later stages. This further suggests that there is a finite limit to the number of stages that can be stacked on a single shaft before it becomes impossible to start the machine. There are several methods which designers have used to resolve the problem of low-speed front stage stall, as described in Sections 2.1.2.1–2.1.2.3.

2.1.2.1   Interstage Bleeds

It is intuitively obvious that if the front stages could be accommodated by allowing them to pass more flow at lower speeds, they would then operate in the unstalled portion of their operating envelope. Since the downstream stages cannot swallow the increased flow, it would have to be diverted overboard. Such an arrangement is variously described as 'blow off' or interstage bleed. If such devices only operate at the lower speeds, some control action will be required to open and close them at the appropriate point in the operating envelope. Similarly, it is obvious that some advantage can be obtained by bleeding air from several stages and by opening and closing these valves in a sequence as opposed to all at once.

The notion of compressing air to an intermediate pressure and then dumping it overboard can only be interpreted as a degradation of the efficiency of the engine during this period of operation. While bleeds have been used in the past as a means of stabilizing compressor operation at low speeds, other methods have also been sought as a means to solve this problem.

2.1.2.2   Variable Position Stators

Another method of dealing with the phenomenon of low-speed compressor stall is to arrange for a stage or blade row whose stagger angle is adjustable. This is shown in diagram form in Figure 2.14.

This feature provides separate control of the performance of each stage. A more detailed description of the origins of stage performance in terms of the commonly accepted non-dimensional terms is provided in Appendix A and the process of stacking stage data to produce an overall compressor map is described in Appendix B. However, for purposes of this discussion, the reader should be aware that the conventional means of presenting the performance of a compressor stage is to non-dimensionalize the data with respect to blade speed, as shown in Figure 2.15.

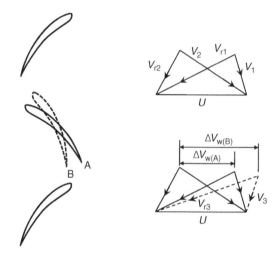

Figure 2.14   Effect of variable position stators on stage performance.

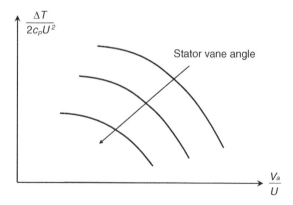

Figure 2.15   Overall stage performance of an axial compressor with variable position stators.

The term V_a/U is called the flow coefficient. This parameter can be shown to be equivalent to

$$\frac{w\sqrt{T}}{P}\frac{\sqrt{T}}{N}$$

which, for a fixed blade speed, is simply the non-dimensional flow rate. Similarly, the term $c_p\Delta T/U^2$ is the temperature coefficient which is equivalent to the parameter $\frac{\Delta T}{T}/\left(\frac{N}{\sqrt{T}}\right)^2$ and $\Delta T/T$ is related to the stage pressure ratio through the thermodynamic relationship:

$$\frac{P_2}{P_1} = \left(1 + \eta_c\frac{\Delta T}{T}\right)^{\frac{\gamma}{\gamma-1}} \tag{2.8}$$

where η_c is the stage efficiency.

It is therefore apparent that a single curve of temperature coefficient versus flow coefficient for a constant efficiency describes a fixed geometry compressor stage. However, if the stage is fitted with variable stators, a family of similar curves is obtained as shown in Figure 2.15.

Careful examination of the velocity diagrams in Figure 2.14 indicates that the orientation of the stator blade dictates the angle of incidence of the flow entering the downstream rotor. The angular position of the adjustable stator affects the absolute velocity V_3 and the relative velocity V_{r3} such that a change in the net whirl velocity ΔV_w is obtained. Since the power absorbed is directly proportional to the change in whirl velocity, the pressure ratio and airflow across the stage are directly affected [5]. Employing the nomenclature commonly used to present a stage characteristic (see Appendix A), the use of a variable stator will provide a two-dimensional stage characteristic as shown in Figure 2.15.

By stacking successive stages, each with a variable stator, a great deal of control is obtained over the overall compressor performance. It is readily observed from Figure 2.14 that a front stage in stall at low speed can be moved to an unstalled state (lower flow rate and lower pressure) by resetting the stage blade angle. Similarly, it is apparent that more stages can be 'stacked' on the same shaft, obtaining higher overall compressor pressure ratios than would otherwise be obtained from a single-spool machine. The General Electric Company was the first to exploit this phenomenon and produced very successful jet engines with as many as 16 stages on a single spool. Figure 2.16 shows a concept drawing of the General Electric J-79 Engine together with a photograph of the engine. This was the first GE engine to be fitted with multiple variable stators on a single axial compressor. As shown in the concept drawing and clearly visible in the photograph, the first six compressor stages of the J-79 had variable stators.

2.1.2.3 Multiple Rotors

Yet another approach to the resolution of the compressor stall problem should be obvious at this point. If we break the compressor into two separate units, each running at different speeds, much of the problem can be resolved and a high overall pressure ratio obtained. Such a solution has the disadvantage of added complexity of shafting; however, it might be argued that such an arrangement is considerably less complicated than multiple stages of variable-stagger stator vanes.

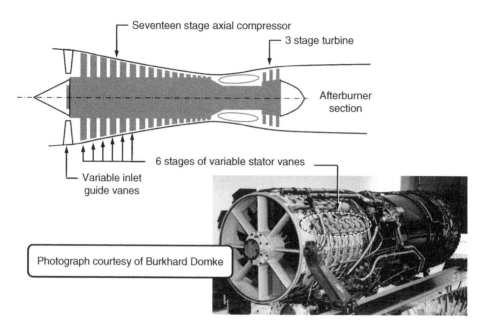

Seventeen stage axial compressor

3 stage turbine

Afterburner section

6 stages of variable stator vanes

Variable inlet guide vanes

Photograph courtesy of Burkhard Domke

Figure 2.16   General Electric turbojet engine concept (Courtesy of Burkhard Domke).

There are numerous examples of multispool engines; this was the standard practice of both Rolls-Royce and Pratt & Whitney and many successful engines have employed various combinations of shaft arrangements and interstage and interspool bleeds. Most high-performance aircraft gas turbine engines are likely to utilize both a multispool layout and interstage bleed together with variable geometry on one or more stages. The Pratt & Whitney JT9D engine, shown in the concept drawing of Figure 2.17, is such an example. Here the two-spool arrangement plus the compressor bleed valve allow for a much less complex variable geometry compressor than the above J-79 example.

2.1.3   Variable Nozzles

The final element of engine performance that requires some description is the use of variable area nozzles with afterburners; the combination is commonly referred to as 'thrust augmentation'. In general, this feature applies only to military engines; however, the Anglo-French Concorde and the Russian TU-144 used thrust augmentation to boost thrust to the levels required for supersonic flight. Both airplanes were an attempt to make supersonic flight commercially viable. Both planes have since been retired without a replacement.

The analysis of the steady-state design performance of a gas turbine at any given rotational speed is an iterative process whereby the equations governing continuity of flow through compressor, turbine, and nozzle are satisfied while maintaining a balance of power between the compressor and turbine. As stated previously, for a fixed area nozzle this results in a singular operating point at each permissible compressor operating speed.

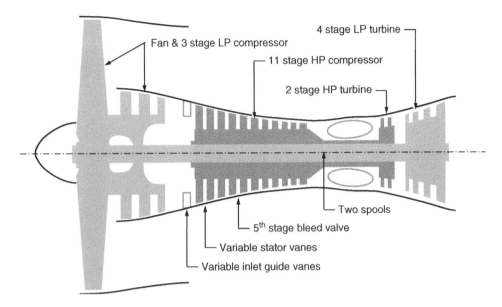

Fan & 3 stage LP compressor

4 stage LP turbine

11 stage HP compressor

2 stage HP turbine

Two spools

5th stage bleed valve

Variable stator vanes

Variable inlet guide vanes

Figure 2.17   Pratt & Whitney JT9D turbofan engine concept.

The locus of these points is commonly referred to as the 'unique operating line' as shown in Figure 2.4. The assumptions which allow this statement to be made are:

1. the nozzle area is fixed; and
2. there is no energy added other than that obtained from the primary combustion chamber.

The addition of an afterburner changes the energy content of the jet pipe flow which, in the absence of an adjustment to the nozzle area, will upset the flow balance and drive the compressor into surge. To understand why this is so we need to consider more carefully how the nozzle and jet pipe interact with the gas generator.

A plot of a typical nozzle characteristic is shown in Figure 2.18.

The parameters describing this plot are non-dimensional flow parameter $w_6\sqrt{T_6}/A_N P_6$ and pressure ratio P_6/P_a, where A_N is nozzle exit area and P_a is ambient pressure. As pressure ratio is increased, the flow parameter continues to rise until the pressure ratio reaches the critical condition shown. At this point, the flow at the nozzle throat has reached sonic conditions and cannot rise further. For the simple convergent nozzle shown in Figure 2.19, sonic conditions occur at the exit plane at which point the pressure remains at the critical value for all values of P_6/P_a which are greater than $(P_6/P_a)_{CRIT}$.

Returning to Figure 2.18 and re-plotting with A_N as a parameter we get a series of curves, each representing a different value of A_N. This is shown on the right-hand side of Figure 2.20 which is adapted from [6].

The left-hand side of Figure 2.20 is a plot of the flow characteristic of a turbine with values of nozzle non-dimensional flow superimposed. These values of nozzle flow are

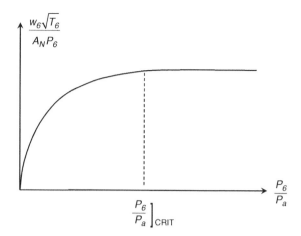

Figure 2.18 Typical nozzle characteristic.

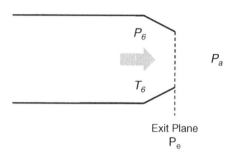

Figure 2.19 Simple convergent nozzle.

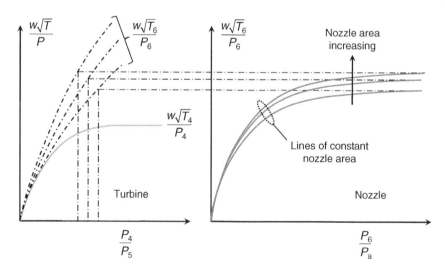

Figure 2.20 Matching a variable area nozzle to a turbine.

obtained from the identity:

$$\frac{w_6\sqrt{T_6}}{P_6} = \frac{w_4\sqrt{T_4}}{P_4}\frac{P_4}{P_5}\sqrt{\frac{T_5}{T_4}}\frac{P_5}{P_6}\sqrt{\frac{T_6}{T_5}}. \tag{2.9}$$

For the case considered here, the tail pipe has no afterburner and the pressure loss is assumed to be negligible. Using this assumption, it can be seen that the identity conveniently casts the nozzle flow parameter in terms entirely determined by the turbine parameters. By graphically matching the choked nozzle condition for three different areas, it can be seen that the larger the area of the nozzle, the lower is the turbine pressure ratio. A second identity allows us to determine the effect that this has on the compressor operating point:

$$\frac{P_6}{P_a} = \frac{P_3}{P_2}\frac{P_4}{P_3}\frac{P_5}{P_4}\frac{P_6}{P_5}\frac{P_2}{P_1}\frac{P_1}{P_a}. \tag{2.10}$$

Assuming sea-level static conditions ($P_a = P_1 = P_2$) and assuming that the tail pipe suffers no loss ($P_5 = P_6$) we can rewrite the pressure identity from Equation 2.10 as follows:

$$\frac{P_3}{P_2} = \frac{P_3}{P_4}\frac{P_4}{P_5}\frac{P_6}{P_a}. \tag{2.11}$$

The term P_3/P_4 represents a pressure loss across the combustion chamber which is of the order 4%. It therefore becomes apparent that, for any value of nozzle pressure ratio, an increase in A_N results in a decrease in turbine pressure ratio and a corresponding decrease in compressor pressure ratio. The effect of nozzle area variation is commonly displayed on the compressor map, as shown in Figure 2.21.

Continuing with the example of the turbojet engine, we shall consider the engine nozzle pressure ratio under a variety of design parameters. The principal design parameters available are the compressor pressure ratio and the turbine inlet temperature.

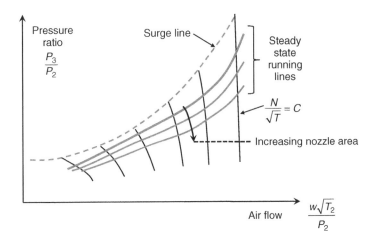

Figure 2.21 Single-spool turbojet compressor operation with a variable area nozzle.

Using the following typical component performance values:

- compressor efficiency: 87%;
- turbine efficiency: 90%;
- engine inlet efficiency: 93%;
- combustor efficiency: 98%;
- combustor pressure loss: 4%; and
- mechanical efficiency: 4%.

We can perform a series of cycle calculations from which the pressure ratio across the nozzle can be calculated. This data is presented in Figure 2.22 for sea-level static conditions.

The figure shows that, for a turbine inlet temperature of 1200 K, the nozzle pressure ratio is about 3.0 for practically all of the design compressor pressure ratios. Assuming a typical military engine with compressor pressure ratio of the order 15–20 and turbine inlet temperatures of the order 1400 K, the nozzle pressure ratio is about 4.0.

Figure 2.23 shows data for this same engine operating at a Mach number M of 0.8 and at an altitude of 15 000 ft. Here the nozzle pressure ratio is of the order 6.0 at the higher compressor pressure ratios. This is due partly to the ram affect of the forward speed on inlet pressure and partly to the reduced ambient back pressure at the nozzle exit for this operating altitude.

Considering the thrust produced by a jet engine, the thrust equation is expressed as:

$$F_N = w \left(V_j - V_{ac} \right) + A_N \left(P_e - P_a \right) \tag{2.12}$$

where F_N is net thrust; w is airflow rate at inlet; V_j is jet velocity at nozzle exit plane; V_{ac} is aircraft forward speed; A_N is nozzle exit area; P_e is pressure at the nozzle exit plane; and P_a is ambient pressure.

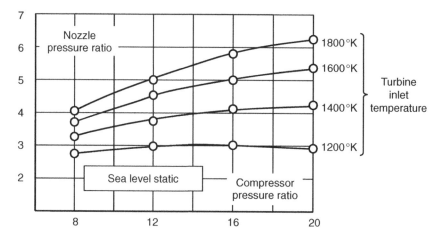

Figure 2.22 Turbojet nozzle pressure ratios at sea-level static conditions.

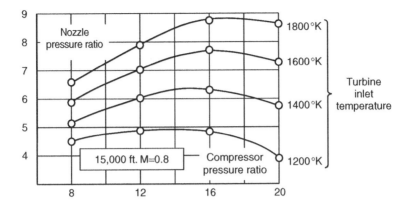

Figure 2.23 Turbojet nozzle pressure ratios at Mach number 0.8 and altitude 15 000 ft.

Note that this equation applies to the conditions of subsonic flight. Under supersonic conditions, the presence of shock waves invalidates the simplifying assumptions which led to Equation 2.13 above.

In light of Equation 2.13, it is apparent that the thrust data obtained in Figures 2.21 and 2.22 comprise a momentum term and a pressure term. Rewriting Equation 2.12 in specific terms, we can estimate the relative magnitudes of these contributions:

$$\frac{F_N}{w} = \left(V_j - V_{ac}\right) + \frac{A_N}{w}\left(P_e - P_a\right). \tag{2.13}$$

Using the conditions associated with a compressor pressure ratio of 16 and a turbine inlet temperature of 1400 K, the specific thrust components can be calculated for the flight condition of 15 000 ft and 0.8 Mach number as follows:

$$\left(V_j - V_{ac}\right) = 270.5 \, \text{N s/kg} \tag{2.14}$$

$$\frac{A_N}{w}\left(P_e - P_a\right) = 237.8 \, \text{N s/kg.} \tag{2.15}$$

It is therefore apparent that the pressure term contributes equally to the momentum term in this case.

It should be noted that the pressure at the nozzle exit is determined by the critical pressure ratio, thus assuring that the nozzle is choked. It should also be noted that the velocity of the flow at the exit of the nozzle is sonic and we can therefore write:

$$V_j = \sqrt{\gamma g_c R T_c} \tag{2.16}$$

where T_c is the static temperature at the critical conditions in the nozzle.

This critical temperature is related to the jet pipe temperature according to the following equation:

$$T_c = \frac{2}{\gamma + 1} T_6. \tag{2.17}$$

The most immediate observation that can be obtained from the above is that we can increase thrust by increasing the jet velocity, which in turn is achieved by increasing the jet pipe temperature. For example, increasing the jet pipe temperature from 1000 K to 1800 K will increase the jet velocity by 34%. This will increase the momentum component accordingly.

If we increase the jet pipe temperature, we must also increase the nozzle area to avoid upsetting the jet pipe pressure. This is because the nozzle choking flow rate parameter $w_6\sqrt{T_6}/A_N P_6 = 0.532$ dictates either a change in w_4 or a change in A_N if P_4 is to remain constant. Since w_4 is actually obtained from the compressor flow rate, the nozzle area must change in order to avoid upsetting the operating point of the compressor.

Increasing the nozzle area and increasing the jet pipe temperature, while holding the pressure in the jet pipe constant, will increase the pressure term of the thrust equation in accordance with the area change. Examining the flow parameter $w_6\sqrt{T_6}/A_N P_6$ and holding P_4 and w_4 constant requires the area to change with the square root of absolute temperature. A change in jet pipe temperature from 1000 K to 1800 K will therefore result in an area change of 34%, exactly as it has affected the velocity noted previously.

The above observation suggests that, if we fit the jet pipe with a burner and appropriate cooling liners, we can raise the average temperature of the flow thereby augmenting the thrust of the engine. A typical arrangement for an afterburner is shown in Figure 2.24. It involves a fuel manifold, fuel nozzles, and some form of gutter or flame holder to ensure that the flame is not blown out while in operation. In addition, a perforated liner is usually employed to cool and protect the jet pipe. Finally, a variable area nozzle is required to maintain the jet pipe pressure at or near the maximum operating point as dictated by the operation of the compressor.

Although we did not cover the idea of a convergent-divergent nozzle in this introductory section, the reader should be aware that such systems are employed for some supersonic military airplanes.

This flight regime results in shock waves in both the inlet and the exhaust. The presence of these shock waves greatly alters the thermodynamic conditions in the inlet and the exhaust and this, in turn, affects the engine. Modern designs employ variable geometry in the inlet and the exhaust in an effort to optimize the aircraft performance; however, this topic is covered in detail in Chapter 6.

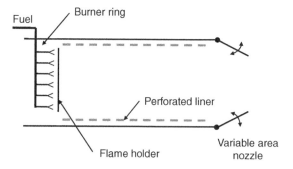

Figure 2.24   Typical afterburner arrangement.

2.2   Concluding Commentary

This chapter provides a brief introduction into the operational and performance principles of the modern gas turbine aircraft propulsion engine in order to set the stage for the primary intent of this book, which is to describe the functional aspects of the various systems and subsystems that are combined with the gas turbine engine to make it a fully integrated functional propulsion system.

References

1. Pratt & Whitney Aircraft, Division of United Aircraft Corporation (1967) The Aircraft Gas Turbine and its Operation.
2. Rolls-Royce (1969–1973) *The Jet Engine*, 3rd edn, Rolls-Royce Ltd.
3. Treager, I.E. (1970–1996) *Aircraft Gas Turbine Engine Technology*, 3rd edn, Glenco/McGraw-Hill.
4. Stone, A. (1957) Effects of Stage Characteristics on Axial-Flow Compressor Performance, ASME Paper 57-A-139.
5. Shepherd, D.G. (1956) *Principles of Turbomachinery*, MacMillan Company.
6. Saravanamuttoo, H.I.H., Rogers, G.F.C. and Cohen, H. (1951–2001) *Gas Turbine Theory*, 5th edn, Pearson Education Ltd.

3

Gas Generator Fuel Control Systems

The core of any gas turbine propulsion system is the gas generator (sometimes referred to as the gas producer) which generates a high-energy gas stream that can be used to provide either thrust or shaft power, dependent upon the application and the specifics of the propulsion system design. Control of the gas generator in terms of fuel flow and, in many applications, the management of airflow through the compressor stages are perhaps some of the most critical functions within the propulsion system package; they are responsible for the safe and effective operation of the engine under both dynamic and steady-state conditions throughout the operational flight envelope.

Typically the gas generator comprises a compressor, combustor, and turbine and, in some cases, a separate low pressure (LP) compressor and turbine with its own concentric shaft, as indicated in the schematic of Figure 3.1.

In many turbofan applications, the LP compressor shaft also drives the fan which generates the majority of the thrust of the engine. Here, the core section of the fan provides some supercharging of the LP compressor stages. The LP turbine may comprise two or more stages.

Generation of thrust and/or shaft power in the various propulsion system concepts involves the absorption of the gas stream energy in one or more turbine stages. Management and control of the resulting thrust or shaft power for the various propulsion system concepts is typically achieved by trimming the gas generator fuel control system in some fashion.

The intent of this chapter is therefore to focus on gas generator control solutions and their functionality as it relates to all aspects of fuel flow metering and compressor air flow management. Subsequent chapters address the control and management of thrust and shaft power engines that are significantly more application-dependent.

3.1 Basic Concepts of the Gas Generator Fuel Control System

The primary requirement of the gas generator fuel control system is to provide the following basic functions: pumping fuel to a sufficiently high pressure (HP) to allow effective

Gas Turbine Propulsion Systems, First Edition. Bernie MacIsaac and Roy Langton.
© 2011 John Wiley & Sons, Ltd. Published 2011 by John Wiley & Sons, Ltd.

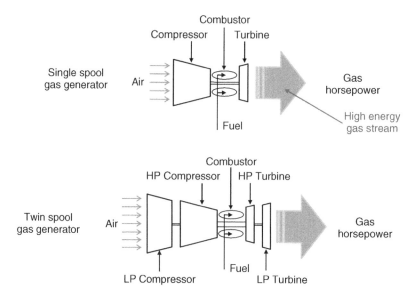

Figure 3.1   Gas generator schematic.

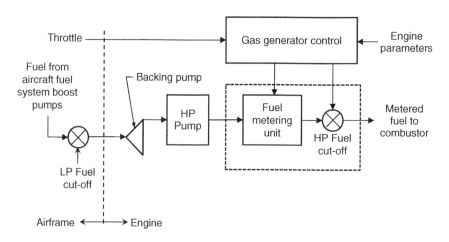

Figure 3.2   Gas generator fuel system overview.

discharge into the combustion chamber, and providing the appropriate fuel metering and control as illustrated in the overview block diagram of Figure 3.2.

This highly simplified depiction shows how the primary fuel control system features interface with the airframe. The airframe is required to supply fuel to the engine within a predetermined range of temperature and pressure. Isolation of the fuel supply from the airframe is via the LP cut-off valve which is part of the aircraft fuel system. The airframe also provides a throttle command to the engine which determines the thrust or power required.

The main task of the gas generator fuel control system is to meter fuel to the combustor that accurately reflects the throttle setting, while maintaining safe operation during transient throttle changes, and to protect the engine from exceeding operational limits associated with internal pressures and temperatures throughout the operational flight envelope.

The modern gas generator fuel system uses a HP fuel pump arrangement comprising a centrifugal backing pump feeding a positive displacement stage in order to develop the high fuel pressures necessary to allow efficient fuel combustion nozzle operation. Since compressor pressure ratios of more than 30:1 are common in many modern engine designs, fuel system delivery pressures can be well in excess of 1000 psi absolute. Typically, the fuel metering unit (FMU) receives commands from the gas generator fuel control which then meters the appropriate amount of fuel to the engine. The FMU also provides a fuel cut-off function to facilitate engine shutdown in response to a flight crew command or following the detection of some unsafe operating condition (e.g., an overspeed event or an engine fire).

The overview of Figure 3.2 does not show the airflow management functions that are typically provided by, and therefore considered to be an integral part of, the fuel control system. This is illustrated in Figure 3.3 which shows a single-spool gas generator with variable compressor geometry in the form of inlet guide vanes (IGVs) and variable stator vanes (VSVs) for several of the LP stages of the compressor. These IGVs and VSVs are positioned as a function of the prevailing engine operating condition, in order to improve compressor performance and surge margins.

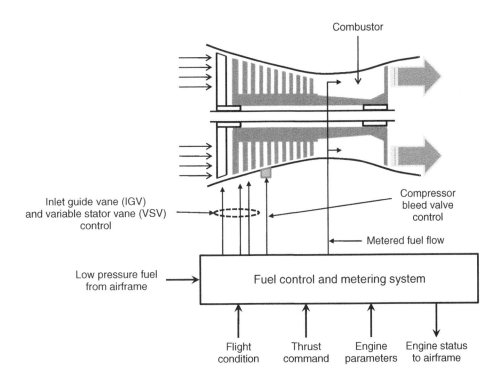

Figure 3.3 Typical gas generator fuel control system tasks.

Bleed valves are also frequently used to improve the compressor surge margin during acceleration from one power setting to another. It is common practice for the fuel control to utilize HP fuel as the medium for IGV and VSV actuators. This concept is often referred to as fuel-draulics. Bleed valves are more typically pneumatically powered.

3.2   Gas Generator Control Modes

This section describes the basic concepts associated with providing safe and appropriate control of fuel flow to the gas generator or 'core' section of the gas turbine engine during transient and steady-state operation.

In order to execute the various control tasks, the fuel control system requires inputs from the throttle, flight condition information, and various engine parameters such as rotational speeds, internal pressures, and temperatures.

One of the most significant challenges associated with the control of aircraft gas turbine engines is associated with the variations in the operational environment. Specifically, these are the variation in air inlet conditions brought about by the effects of altitude, temperature, and airspeed. As a result, the fuel flow required to operate the engine at any given rotational speed varies considerably with changes in the prevailing operating environment.

To address this problem, the non-dimensional principles described in Chapter 2 are used to describe engine performance and to establish control mode strategies from which fuel control system design methods can be established.

The fuel control community often uses these non-dimensional parameters such as fuel flow and speed in what is termed a corrected form. For example,

Fuel flow (w_{Fe}) corrected to the engine inlet conditions becomes: $\dfrac{w_{Fe}}{\delta_1 \sqrt{\theta_1}}$

where δ_1 is actual inlet total pressure in psi absolute/14.7 and θ_1 is actual total air inlet temperature/518.7 in degrees Rankine terms.

Similarly, corrected engine speed becomes $N/\sqrt{\theta_1}$.

In each case the subscript represents the engine stage location: 1 is the engine inlet, 2 the compressor inlet, and so on as defined in Chapter 2, Figure 2.2. The reader should be aware that these definitions can change depending upon the engine configuration and/or engine manufacturer preferences.

Using these corrected parameters, the variations in the engine steady running line and other critical performance features such as surge or flame-out boundary come together as essentially single curves as shown in Figure 3.4. From these graphs, the actual values of engine speed and fuel flow are readily obtained by inserting the values of θ and δ for the actual operating condition being considered.

From the figure it can be seen that, in order to change engine power settings and avoid crossing the critical boundaries associated with compressor performance (acceleration limiting) or combustor flame-out (deceleration limiting), fuel flow must be controlled in some manner. In addition, the controlling function must recognize the dependence of ambient conditions expressed through the corrected forms of the parameters used.

Reference to Figure 3.4 suggests that to change the power level of a gas generator is to control rotor speed. While not universally true, the vast majority of jet engines are in

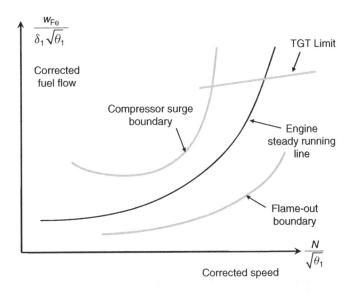

Figure 3.4   Engine performance showing surge and flame-out boundaries.

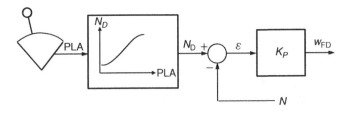

Figure 3.5   Simple proportional control logic.

fact controlled by speed. This suggests the use of simple proportional control, as shown in Figure 3.5.

In the figure, power lever angle (PLA) is mapped into a gas generator speed demand. This speed demand is then compared to the actual engine speed forming an error term ε which is then multiplied by a constant of proportionality which converts it into a fuel flow demand w_{FD}. This demand signal is then provided to a fuel metering system which arranges for this flow rate to be supplied to the engine. This is the logic of the simple droop governor. The term is derived from the fact that a finite error between the speed demand and actual speed must exist in order to generate the associated fuel flow required; the actual engine speed must 'droop' below the demand speed for the concept to work.

The form depicted in Figure 3.5 cannot work for several reasons, which are embedded in Figure 3.4 above. The most formidable obstacle for the droop governor is the compressor surge boundary. If, for example, the engine were running at a low speed such as idle and the PLA was to be advanced quickly to a high power setting, the logic of Figure 3.5 would dictate a new fuel flow demand which would immediately drive the compressor

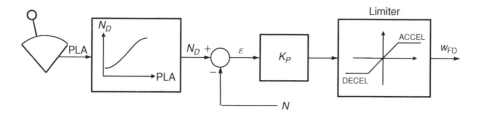

Figure 3.6   Proportional control with limiter.

into surge. A limit must therefore be imposed on the amount of fuel that can be demanded. This suggests a modification to the logic shown in Figure 3.5 to that of Figure 3.6.

The presence of the limit function forces the demand for fuel to stay within the bounds shown. The limits cannot be simple constants, however; they must be adjusted dynamically throughout the transient so that the constraints imposed by Figure 3.4 are not violated and, at the same time, ensure a smooth acceleration to the higher speed required. Furthermore, the schedule of limits must be corrected for ambient conditions. To accomplish this, there are a number of options which will be discussed in the following section.

3.2.1   Fuel Schedule Definition

A common (and the most obvious) fuel-limiting schedule is to map the available margin shown in Figure 3.4 such that a limit function is formed as:

$$\frac{w_F}{P_1\sqrt{T_1}} = f(N, T_1). \tag{3.1}$$

Since $\sqrt{T_1}$ is a relatively weak relationship affecting allowable fuel, the limit can be simplified to

$$\frac{w_F}{P_1} = f(N, T_1). \tag{3.2}$$

This gives up a small amount of the available margin in the interests of simplicity; however, P_1 is usually replaced by compressor delivery pressure P_3 in the form:

$$\frac{w_F}{P_3\sqrt{T_1}} = f(N, T_1). \tag{3.3}$$

This parameter is easily obtained by multiplying the fuel flow in non-dimensional form by pressure ratio as follows:

$$\frac{w_F}{P_1\sqrt{T_1}}\frac{P_1}{P_3} = \frac{w_F}{P_3\sqrt{T_1}}. \tag{3.4}$$

Again, the temperature term is commonly neglected.

It is apparent from the above that any combination of non-dimensional terms could be used provided they contained the fuel term. One such control limiter was developed by Rolls-Royce in the 1960s for application to their large commercial engines. This limiting schedule employs compressor pressure ratio, spool speed, and engine inlet temperature as follows:

$$\frac{w_F}{N} = f\left(\frac{P_3}{P_1}, P_1\right). \tag{3.5}$$

This is obtained by combining the non-dimensional fuel flow with non-dimensional speed as follows:

$$\frac{w_F}{P_1\sqrt{T_1}}\frac{\sqrt{T_1}}{N} = \frac{w_F}{P_1 N}. \tag{3.6}$$

By plotting w_F/N against P_3/P_1, we obtain a new limiting function which can be used for both acceleration and deceleration. To implement such a schedule, it becomes necessary to multiply the parameter w_F/N by N to obtain the limiting fuel flow. This was accomplished at Rolls-Royce by employing a fuel metering valve with a pressure drop regulator connected to a fly weight device via the engine accessory gearbox. The pressure regulator setting is therefore proportional to N^2 and flow through the metering valve is therefore proportional to metering valve position, that is, (flow area) $\times \sqrt{N^2}$.

The popularity of the limiting functions described in Equation 3.3 above was driven by the relative simplicity with which the measurement of compressor delivery pressure and speed could be represented in a hydromechanical control system.

These devices are essentially mechanical analogs of control equations. Thus, their computing medium is force. Speed is converted into a force by a simple flyweight device $F = KN^2$. Pressure requires no elaborate transduction at all since its units are force per unit area.

The choice of compressor delivery pressure is obvious since it is directly related to compressor outlet conditions. The essential compromise leads to the selection of a simple limiting function which provides reasonable transient response at the lowest possible cost and complexity.

The emergence of electronic controls, particularly digital electronics in the 1980s, afforded the industry an opportunity to re-examine the entire functionality of the fuel control. During this re-evaluation, it became apparent that the measurement of pressure was expensive if the requisite accuracy and reliability was to be achieved. The issue of formulating a new fuel limit therefore came under scrutiny.

Returning to the non-dimensional parameters, it was observed that acceleration rate (or N-dot) was a possibility. The rotor acceleration rate is related to the excess torque on the engine rotor during a transient that is, in turn, related to the increase in fuel flow. This parameter has the non-dimensional form $\dot{N}/P_1$ which unfortunately still requires the measurement of pressure. However, by combining non-dimensional groups, it can be presented as follows:

$$\frac{\dot{N}N}{w_F} = \frac{\dot{N}}{P_1}\frac{N}{\sqrt{T_1}}\frac{P\sqrt{T}_1}{w_F}. \tag{3.7}$$

This parameter requires the measurement of speed and fuel flow. Speed is considered an easy measurement to make; since fuel flow must be controlled by a hydromechanical device whose input is electronic, a signal is readily obtained of fuel flow. As will be discussed in Section 2.3.2, the implementation of an N-dot control concept introduces new stability problems in governance of rotor speed.

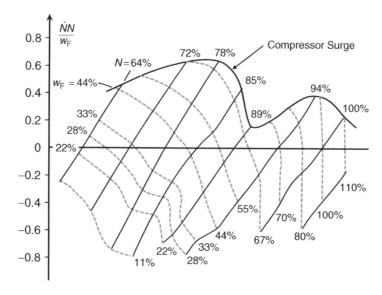

Figure 3.7   Acceleration characteristics of a typical gas turbine engine.

Returning to Figure 3.4, the presentation of a compressor surge boundary in terms of fuel flow as a function of rotor speed must be interpreted in the context of the operation of the compressor during a transient. As discussed in Section 2.1 and depicted in Figure 2.6, the initial transient path from a given steady-state point is, for all practical purposes, along the corresponding speed line in the direction of surge. The point on the surge line, at that speed, thus represents the maximum permissible fuel flow rate before the compressor surges. This also represents the maximum level of excess torque that can be applied to the shaft and therefore the maximum possible rotor acceleration that can be generated.

Modern computers permit us to simulate an engine at the component level (see Appendix C). By fixing the rotor speed and allowing incremental overfueling, it is possible to map all possible levels of overfueling and present this data in terms of the fuel parameter that might be used to limit fuel throughout the transient. An example of such a map is shown in Figure 3.7.

Using these data, together with related engine data (e.g., P_3) any fuel-limiting schedule can be constructed. The surge values of $\dot{N}/P_1$, $\dot{N}N/w_F$, and w_F/P_3 are shown in Figure 3.8 as a function of $N/\sqrt{T_1}$. It is apparent from this figure that any schedule can be constructed from knowledge of the steady-state value of the parameter and its corresponding value at the point of compressor surge.

A similar set of data can be constructed for a deceleration limit where the limiting thermodynamic function is flame stability within the combustor. A typical flame stability boundary is shown in Figure 3.9 [1].

During a rapid deceleration, the engine pressure ratio falls below the steady-state operating line along a constant speed line on the compressor. In this region of operation, the compressor speed line is approximately vertical showing little variation in air flow rate with pressure ratio. There is therefore a rapid rise in air/fuel ratio and the combustor

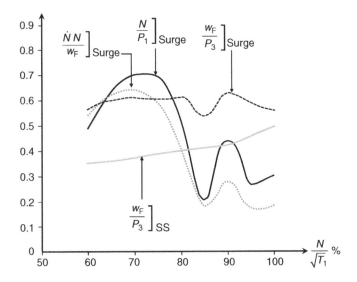

Figure 3.8   Compressor surge boundaries for several common non-dimensional fuel parameters.

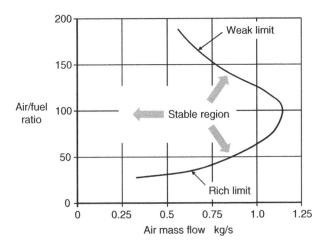

Figure 3.9   Typical combustor stability limits.

operating point migrates toward its region of lean extinction. A limit must therefore be imposed on fuel flow to prevent engine flame-out.

3.2.2   Overall Gas Generator Control Logic

Control of a gas generator must be carried out within the hard limits of compressor surge, top temperature limits, and combustor flame out as indicated in Figure 3.4. It thus represents an extremely non-linear system with control passing from one limit to

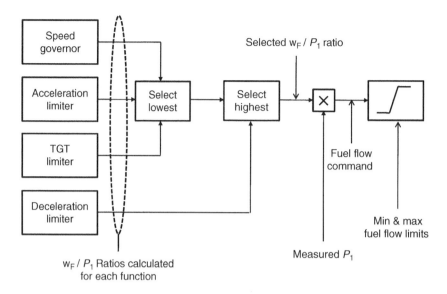

Figure 3.10 Conceptual block diagram for the w_F/P_1 control mode.

another, depending on the situation. Let us therefore consider the complete picture of a gas generator control from the perspective of a logic block diagram. For this purpose we will use the so-called control mode:

$$\frac{w_F}{P_1} = f(N, T_1).$$

Here all control and limiting functions are expressed in the form of the ratio of fuel flow to engine inlet pressure. Logic is then used to select the appropriate controlling function. This is illustrated in the conceptual block diagram of Figure 3.10. While the above control mode is one of the common techniques for managing the acceleration, deceleration, and control of aircraft gas turbines, there are many other approaches in use.

For example, Figure 3.10 can easily represent the popular control mode:

$$\frac{w_F}{P_3} = f(N, T_1).$$

Here P_1 has been replaced by compressor discharge pressure P_3 (this parameter is sometimes designated P_C, P_{CD}, or CDP). The term P_B is sometimes used by fuel control specialists; in this case, P_B means 'burner pressure' which is essentially the same as CDP.

3.2.3   Speed Governing with Acceleration and Deceleration Limiting

When operating in steady state, the gas generator is typically controlled by a simple proportional (or droop) speed governor where speed is determined by the throttle position or PLA.

From the above discussion and Figure 3.10, when transitioning from one steady-state speed to another the engine must be protected from compressor surge during acceleration, flame-out during deceleration, and from exceeding turbine temperature limits.

A more detailed acceleration, deceleration, and governing control schematic is presented in Figure 3.11 based on the $w_F/P_1 = f(N, T_1)$ control mode.

During an acceleration, the output from the governor is prevented from exceeding the acceleration schedule by the select low logic. Similarly, during deceleration the deceleration schedule will prevail courtesy of the select high logic.

Figure 3.12 shows the locus of engine operation during transient throttle changes from idle to maximum speed. Referring to the figure, assume that the engine is operating initially at idle speed (point A) on the engine steady running line. Consider now a sudden change in throttle setting commanding a speed change from idle to maximum. As fuel is increased causing the engine to accelerate, the fuel control must limit the amount of overfueling to prevent compressor surge.

As the engine accelerates, the fuel flow to the engine is scheduled to follow the acceleration limit until it reaches the maximum speed governor droop line. From this point, any further increase in speed is now accompanied by a corresponding reduction in fuel flow until the engine steady running line is reached at point B. Similarly, during a deceleration from maximum to idle, fuel flow is maintained above the flame-out boundary by the fuel control's deceleration limit. Upon reaching the idle speed droop line, the governor takes over and brings the engine back to point A.

As mentioned previously, the governor concept illustrated in the figure is a 'droop governor' which means that an error between the speed set by the throttle and the actual running speed must exist in steady state in order to generate a fuel flow command to the fuel metering system.

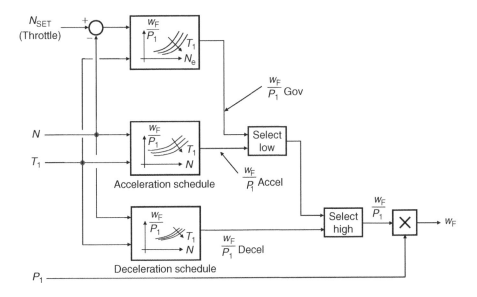

Figure 3.11   Acceleration, deceleration, and speed governing control.

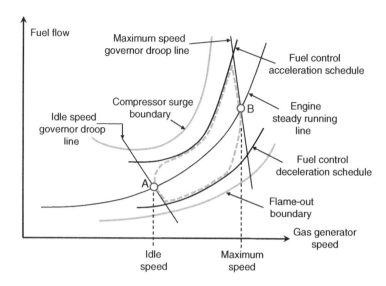

Figure 3.12 Transient operation between idle and maximum speeds.

Since the output of the speed governor is typically in ratio units (i.e. fuel flow divided by either inlet pressure or compressor discharge pressure) there is an inherent compensation for altitude and forward speed within the control loop. Variations in inlet temperature, if not accounted for, will result in a shift in the engine running speed for any given throttle setting. It is therefore common practice to include inlet total temperature within the governor control algorithm in order to provide consistent speed operation as a function of throttle setting. One such approach is shown in the control schematic of Figure 3.11 above, where the speed error is biased by T_1.

3.2.3.1   Governor Response and Stability

To cover this topic adequately involves reference to classical feedback control techniques. Readers not familiar with this topic should refer to Appendix D which contains a brief primer on the subject; a more detailed treatment can be found in [2], which is recommended as an easy-to-read introduction to the subject.

A fast and effective way to analyze the dynamic performance of the speed governor control loop is to use the small perturbation method: the dynamic elements are linearized about a specific operating point and, as the term implies, the results are valid for small excursions about that point. This allows the systems engineer to get a good feel for the dynamics of the speed governor in terms of bandwidth and stability margins; however, this should ultimately be verified via a full-range dynamic simulation of the engine and fuel control system using digital modeling techniques.

Let us first analyze the performance of a speed governor which employs the $w_F/P_1 = f(N, T_1)$ control mode using the small perturbation method. The control loop for this type of governor can be represented by the simplified block diagram of Figure 3.13 which shows all of the major elements involved.

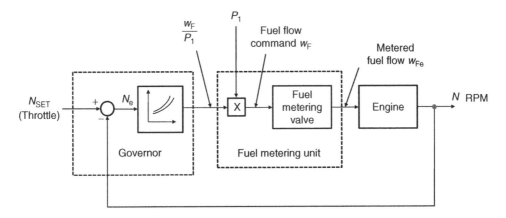

Figure 3.13   Gas generator governor block diagram: $w_F/P_1 = f(N, T_1)$ control mode.

The dynamics associated with the governor and FMU are fast relative to the engine and remain fairly consistent over the operational flight envelope. The governor block can be represented by a simple gain term equal to the slope of the speed error ratio characteristic at the operating point under consideration, together with a linear lag of 10 ms. The fuel metering valve can be adequately represented by a linear lag of 20 ms for the purpose of our linear analysis.

The engine block in the figure can be represented by the following transfer function for any steady-state condition and is valid for small perturbations in fuel flow about the operating point being considered:

$$\frac{\Delta N}{\Delta w_{Fe}} = \frac{K_e(e^{-as})}{(1 + \tau_c s)(1 + \tau_e s)}. \tag{3.8}$$

The gain term K_e in the numerator is referred to as the engine gain and is a measure of the sensitivity of the engine, in terms of shaft speed, to changes in fuel flow. This parameter, which is the inverse of the slope of the fuel flow versus engine speed steady-running line, has a low value at high power and increases as power is reduced to idle.

The term e^{-as} in the numerator is a transport delay which represents the time delay between a fuel flow change at the combustion nozzle and the resulting change in gas state appearing at the throat of the turbine. This term, which is related to the combustion chamber dimensions and the airflow through the engine, is very small but can become significant at low power and high altitudes. It is therefore prudent to make some allowance for this parameter in any governor dynamic analysis.

As discussed in Chapter 2, the engine speed response is dominated by the rotor inertia which must be accelerated by the excess torque generated by the turbine above that required by the compressor until a new steady state is reached and the torques are again in balance. A time constant (τ_e) of the order 0.3–0.5 s is typical for the high-power condition. Interestingly, it is fairly independent of the size of the engine because the torque imbalance scales reasonably well with the increased rotor inertia associated with a bigger engine. However, for a given engine, this time constant becomes much longer at higher altitudes or as throttle setting is reduced toward idle.

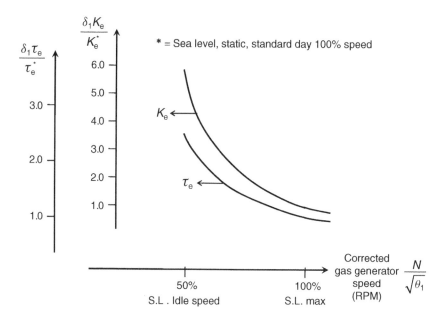

Figure 3.14   Engine gain and time constant variation.

The first-order lag represented by the time constant τ_c represents the dynamics of the combustion process and is also very small compared to the primary engine time constant τ_e. For the purpose of simplifying our analysis we will neglect the combustion dynamics; these terms should however be considered in any rigorous analysis, particularly for operation at high altitudes where combustion dynamics can become more significant.

The effects of the engine operating environment on both the engine gain and the associated primary time constant vary considerably. Figure 3.14 shows how both the engine gain K_e and engine time constant τ_e vary with corrected speed relative to their nominal sea-level maximum-speed values.

When the engine is operating at sea level, static standard day conditions, both δ_1 and $\sqrt{\theta_1}$ are unity. As engine speed is reduced to idle, the value of engine gain K_e increases by a factor of about 5 while the engine time constant increases by a factor of > 2.

Over a typical commercial aircraft flight envelope, δ_1 can vary from about 0.25 to 1.2 while $\sqrt{\theta_1}$ varies over a much smaller range from about 0.9 to 1.05. The values of the engine time constant τ_e and engine gain K_e relative to τ_e^* and K_e^*, respectively can be readily obtained from Figure 3.14 by inserting the appropriate values of δ_1 and $\sqrt{\theta_1}$ for the flight condition being analyzed.

Since the engine dynamics are much slower than the dynamics of the fuel control and metering equipment, the speed governor is relatively easy to design as a closed loop control system. Here the engine time constant is at its fastest at about 0.5 s while the control and metering dynamics are more than an order of magnitude faster. In closed loop control systems, it is good practice to have the dynamics of the process being controlled separated from the controller dynamics by a decade or more. (A decade in this case is defined as a factor of 10 in frequency response capability. In simple terms, this

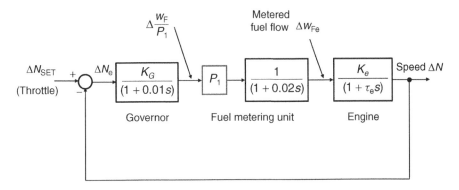

Figure 3.15   Linearized governor block diagram: $w_F/P_1 = f(N, T_1)$ control mode.

means that the controller is 10 times faster than the engine.) The fact that the engine becomes slower with altitude and reducing power only helps the situation, since the control and metering dynamics do not change significantly with operating condition.

We can now construct a block diagram for the governor using transfer functions representing a linearized model that is valid for small perturbations about any selected operating condition: this is shown in Figure 3.15.

In the diagram the term K_G is the linearized governor gain in ratios per unit speed error for the operating point being considered. For the $w_F/P_1 = f(N, T_1)$ control mode, the multiplier becomes a simple gain term since P_1 is constant for any given operating condition. The dynamic terms for the governor and FMU can be assumed to remain the same for all operating conditions, that is, linear lags with time constants of 10 and 20 ms respectively.

The engine gain K_e and time constant τ_e will have to be selected for the flight condition and throttle setting under study.

From Figure 3.15 we can express the open-loop transfer function for this governor as follows:

$$\frac{\Delta N_e}{\Delta N} = \frac{K_G P_1 K_e}{(1 + 0.01s)(1 + 0.02s)(1 + \tau_e s)}. \tag{3.9}$$

It is a relatively simple process to substitute $s = j\omega$ in the above expression and to generate a Bode Diagram (frequency response plot) in order to establish a loop gain that will give the required stability margins.

An alternative approach, and one that provides a better understanding of the dynamic behavior of linear systems such as our speed governor, uses the frequency domain to show graphically where the roots of the open-loop transfer function are located and how the closed-loop roots move within this domain as the loop gain is increased.

This is called the root locus technique and is extremely effective in the study of linear closed-loop systems (see Appendix D for an introduction to this concept).

This technique is developed from the characteristic equation for any closed loop system, which is:

$$1 + \text{the open-loop transfer function} = 0. \tag{3.10}$$

This equation defines the dynamic stability characteristics of the system. It also describes the condition for marginal stability where sustained oscillations will occur if the gain around the loop is unity and the phase lag around the loop is 180°.

The characteristic equation can also be expressed as follows:

$$(\text{Product of all the elements around the loop}) = -1 = |1.0| \angle 180° \qquad (3.11)$$

In a typical example consider a system characteristic equation of the following form:

$$\frac{K(s + z_1)(s + z_2)}{(s + p_1)(s + p_2)(s + p_3)} = |1.0| \angle 180° \qquad (3.12)$$

The z values in the numerator are called zeros since the expression tends to zero when s is set to any of the z values. In the denominator, the p values are referred to as poles since the expression tends to infinity when s is set to any of the p values.

It follows that if we can define a locus of all of the points in the frequency domain (the s plane) in which the sum of all of the vector angles from all of the zeros to any point on the locus minus the sum of all of the vector angles from all of the poles to the same point on the locus is equal to 180°, we will be able to see how the closed-loop roots of the system move in the s plane as the value of K varies from zero to infinity.

The value of K at any point on the locus is simply the product of the vector lengths (moduli) from the poles to the point on the locus divided by the product of the vector lengths from the zeros to the same point.

The location of these root loci in the s plane is a simple process supported by a number of simple rules that allow the analyst to quickly sketch the loci (see Appendix D).

The number of loci equals the number of poles. These loci travel to one of the zeros or, in the case where there are no zeros, to infinity along predetermined asymptotes.

We can apply the root locus concept to our speed governor by defining the system characteristic equation as follows:

$$\frac{K_L}{(s + 1/\tau_e)(s + 50)(s + 100)} = -1 \qquad (3.13)$$

where K_L is the loop gain $= (1/\tau_e)(50)(100)K_G P_1 K_e$.

We can now develop a frequency domain plot of our system showing the location of the three system poles, as depicted in Figure 3.16.

In the frequency domain, $s = \alpha + j\omega$. The y axis defines the $j\omega$ term which represents frequency in the real world while the x axis defines α which represents the rate of decay (or growth) of oscillation according to the real value in the plane. The more negative the location, the more rapid the decay. Points in the right half plane imply unstable situations where oscillations grow rather than decay. Points on the y axis represent sustained oscillations at a frequency corresponding to the value of $j\omega$.

In our governor example, there are three poles: one at -100, one at -50, and a third representing the primary engine lag close to the origin.

Root locus theory predicts that there will be a locus between the engine pole at $-1/\tau_e$ and the fuel metering pole at -50, which splits into two loci at about $\alpha = -20$ and then moves toward the asymptotes as shown.

Since the pole associated with the engine lag τ_e has a maximum value of about -2.0 (representing a lag of 0.5 s), all other values for this parameter will have a pole closer to

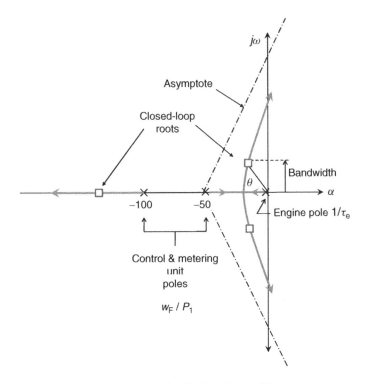

Figure 3.16 Frequency domain plot for the w_F/P_1 speed governor.

the origin. Clearly, any variation in the engine primary time constant will not significantly change the trajectory of the loci shown on Figure 3.16.

The loci that impact the stability of the system are the two that cross the $j\omega$ axis at about 70 rad/s. This establishes the loop gain for marginal stability. The third locus starts at the -100 pole and continues along the negative x axis toward minus infinity and will have little impact on the system response or stability.

In selecting the closed-loop gain required, and hence the position of the closed-loop roots on the loci, we must aim for a well-damped closed-loop behavior. Since the cosine of the angle θ on the graph is equal to the damping ratio, we have chosen an angle of about $45°$. This represents a damping ratio of 0.707, yielding a response with no overshoot following a step change in input. Governor bandwidth is about 20 rad/s (about 3 Hz) implying a fairly fast and well-behaved response to throttle changes.

It is important to note that, in neglecting some of the faster dynamics in the system, results obtained are somewhat optimistic (particularly when considering the higher frequencies). For example, if we add an additional linear lag of 10 ms to the system the root locus will change as indicated in Figure 3.17. Note that the critical locus bends more toward the right half plane, crossing the y axis at a much lower frequency than in our original simplified governor representation. Since we chose a well-damped solution for our loop gain, the location of these two closed-loop roots does not change significantly.

Having located the desired position of the closed-loop roots in the s plane, we can determine the corresponding value of governor gain K_G from Equation 3.9 above for the

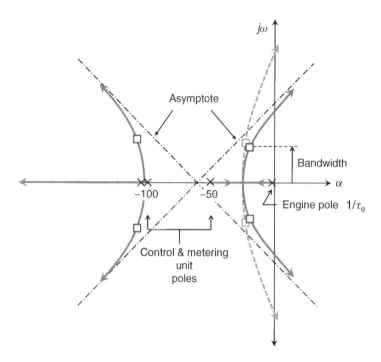

Figure 3.17   The effect of an additional lag on the root loci.

loop gain. Note that the loop gain varies with both the engine gain K_e and the primary engine time constant τ_e; it will therefore be necessary to modify the value of K_G as a function of engine speed to maintain consistent governor performance. The P_1 multiplier provides a natural gain compensation for the effects of altitude and forward speed.

In the schematic of our example shown in Figure 3.6, the governor output is a function of speed error and inlet temperature. The slope of w_F/P_1 plotted against speed error characteristic is the effective governor gain at the steady-state operating point. This shows that, for low-throttle settings where the required w_F/P_1 is relatively low, the slope of the curve is lower than at the high-throttle settings where the required w_F/P_1 ratio is much higher. The temperature modulation compensates for the shift in the engine steady running line with inlet total temperature to provide a consistent speed versus throttle setting relationship.

The gas generator governor described above is based on the $w_F/P_1 = f(N, T_1)$ control mode. The alternative control mode $w_F/P_3 = f(N, T_1)$ described in Section 3.2.1 above presents more of a challenge to the speed governor designer due to the fact that the compressor discharge pressure P_3 is an engine variable whose dynamic behavior has a significant impact on governor stability. The P_3 multiplication process also provides a positive (destabilizing) feedback from the engine to the FMU.

The speed governor block diagram for this type of governor is shown in Figure 3.18. There are now two outputs from the engine to the fuel control system – speed and compressor discharge pressure – which must be taken into account in any dynamic analysis.

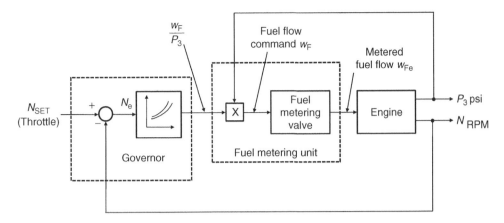

Figure 3.18 Gas generator governor control loop: w_F/P_3 control mode.

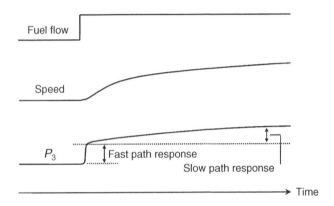

Figure 3.19 Engine response to a step change in fuel flow.

As mentioned previously this feedback is positive, meaning that an increase in P_3 will cause an increase in fuel flow to the engine. For the analysis we must now develop the dynamic relationship between P_3 and fuel flow.

Typical time responses of speed and compressor discharge pressure for a small step change in fuel flow are shown in Figure 3.19. As shown, the engine speed response is slowest and dominated by the primary engine time constant τ_e. Compressor discharge pressure P_3 has a two-stage response, often designated the fast-path and slow-path responses.

Initially there is a sudden increase in pressure due to the addition of fuel into the combustor; this is the fast path response. This is followed by a further increase in pressure as speed and airflow increase as the engine accelerates to the new steady-state condition. (It should also be noted that turbine gas temperature or TGT response also shows a fast-path/slow-path type of response; however, the slow path causes a reduction in temperature as airflow through the machine increases.)

We can develop the small perturbation P_3 response from the following equation which represents the sum of the fast-path and slow-path contributions:

$$\Delta P_3 = \frac{\partial P_3}{\partial w_{Fe}} \Delta w_{Fe} + \frac{\partial P_3}{\partial N} \Delta N. \tag{3.14}$$

For the FMU we must linearize the multiplier about the operating condition since the P_3 input is now an engine output variable and not, as in the case of P_1, a constant determined from outside the control loop. We can therefore define the fuel flow command to the fuel metering valve via the equation:

$$\Delta w_F = \frac{\partial w_F}{\partial R} \Delta R + \frac{\partial w_F}{\partial P_3} \Delta P_3 \tag{3.15}$$

where R is the ratio w_F/P_3 coming from the governor.

The value of the partial derivatives in the above equation can be determined from the following equations:

$$w_F = R P_3, \tag{3.16}$$

$$\frac{\partial w_F}{\partial R} = P_{3O}, \tag{3.17}$$

and

$$\frac{\partial w_F}{\partial P_3} = R_O. \tag{3.18}$$

The values of P_{3O} and R_O are the nominal values of P_3 and R for the operating condition being considered. We can now construct a small signal block diagram for the system, as depicted in Figure 3.20.

The dynamics associated with the fuel metering hardware and the engine speed responses remain the same as previously described. In order to analyze the dynamic

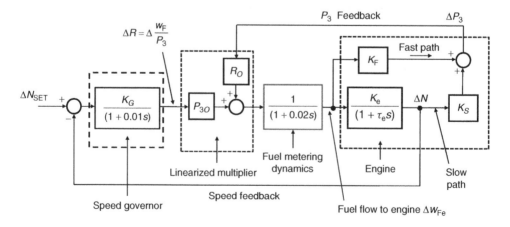

Figure 3.20   Small signal governor block diagram: w_F/P_3 control mode.

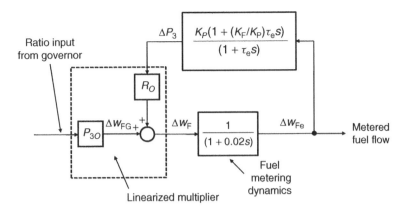

Figure 3.21 P_3 feedback loop dynamics for small perturbations.

performance of this system, we need to rationalize the above block diagram by defining the P_3 feedback loop separately as follows:

$$\Delta P_3 = K_F \Delta w_{Fe} + \left(\frac{K_e}{(1 + \tau_e s)}\right) K_S \Delta w_{Fe}. \tag{3.19}$$

Rationalizing the above expression and letting $K_P = K_F + K_e K_S$ we obtain:

$$\frac{\Delta P_3}{\Delta w_{Fe}} = \frac{K_P[1 + (K_F/K_P)\tau_e s]}{(1 + \tau_e s)}. \tag{3.20}$$

From Figure 3.21, which shows the P_3 feedback loop for small perturbations, it can be seen that if the product $R_O K_P$ is greater than 1.0 then this loop will be unstable, having a root in the right half of the s plane.

Without getting into a detailed analysis of this loop, it suffices to say that the effect of this inner P_3 loop is to make the engine and the metering dynamics slower. Figure 3.22 shows a root locus plot of a typical governor of this type, showing how the governor itself will become stable provided that the loop gain is large enough for the root to cross the y axis into the left half plane. Also, since the fuel metering pole moves to the right, the breakaway point on the locus will be nearer the origin. As a result, the achievable loop gain (and governor bandwidth) will be less than for the previously discussed governor. A good aspect of this type of governor is that the P_3 feedback provides gain compensation for both throttle setting and flight condition.

3.2.3.2   N-Dot Acceleration and Deceleration Limiting

In the above discussion, fuel flow to the engine is determined by a droop speed governor that operates within the constraints of acceleration and deceleration limits. These limits are scheduled as a function of the prevailing operating condition.

With the development of digital control systems, the possibility of other means of controlling fuel during acceleration and deceleration were examined. An alternative to the traditional w_F/P_3 control mode, which has been adopted in some modern aircraft

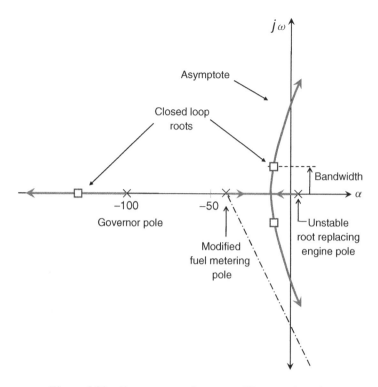

Figure 3.22   Governor root locus: w_F/P_3 control mode.

propulsion applications, is referred to as the N-dot acceleration technique. The basic idea is to control the actual rate-of-change of shaft speed directly as a function of the engine operating conditions. One such N-dot approach is referred to as the slave datum technique which is illustrated in Figure 3.23. Here, instead of computing fuel flow limits to control acceleration, the N-dot controller calculates a limiting engine acceleration or deceleration rate according to the prevailing operating condition. Throttle commands to the N-dot controller are prevented from exceeding the computed N-dot limits using select logic gates as before.

In its earliest form, simple rate limiting was found to be adequate for situations where rapid response was not a major design driver. However, regardless of how the limit is selected, the requirement to avoid compressor surge on acceleration and combustor flame-out on deceleration remains. N-dot can be compensated for ambient conditions as $\dot{N}/P_1$ which requires the measurement of engine inlet pressure or, alternatively, using $\dot{N}N/w_F$ which requires feedback from the engine in the form of rotor speed and fuel flow. In either case, the available surge margin shown in Figures 3.7 and 3.8 apply to the engine in question.

The output from the N-dot controller shown in Figure 3.23 is input to an integrator whose output is a speed command to a conventional droop-type speed governor.

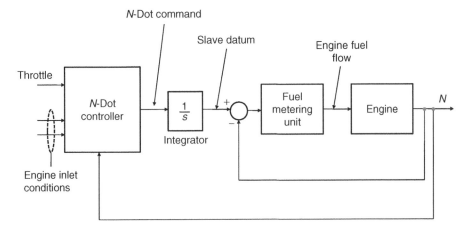

Figure 3.23 *N*-dot acceleration control example.

Since we now have an integrator in the outer speed control loop, a throttle command to the engine will always correspond to a specific engine speed for all flight condition. In other words, there will be no speed droop since, for any finite speed error, the integrator output will continue to change until the throttle speed setting matches the actual engine speed. This type of governor is referred to as an isochronous governor.

The inner-loop speed governor will have a speed droop as before, but this governor loop is there only as a means to allow the engine speed to follow the slave datum.

From a response and stability perspective, however, the presence of an integrator in the outer speed-control loop requires some dynamic compensation features in order to ensure good throttle response and speed governing control-loop stability.

The root locus plot of Figure 3.24 shows how the outer-loop governor performance is improved by the use of lead-lag compensation.

It should be noted that this plot has a different scale than previous root locus diagrams for the purpose of improving clarity of the loci near the origin. In this plot, the closed-loop inner-loop droop-governor roots become open-loop roots in the outer-loop speed control loop. The pole at the origin is the integrator whose output is the slave datum for the inner governor to follow; the zero at -10 together with the pole at -100 represent the lead lag term.

The two oscillatory roots of the outer speed-control loop move along the loci from the droop-governor poles to cross the $j\omega$ axis at about 40 rad/s. The remaining two closed-loop roots are located on the negative real axis; the first is formed on the locus between the integrator pole and the lead-lag zero and the second on the locus between the lead-lag pole and minus infinity.

The performance of the outer speed-control loop is dominated by the real root near the origin which represents a time constant of about 0.1 s and the complex conjugate roots which show a moderately damped second-order term with a natural frequency of about 20 rad/s.

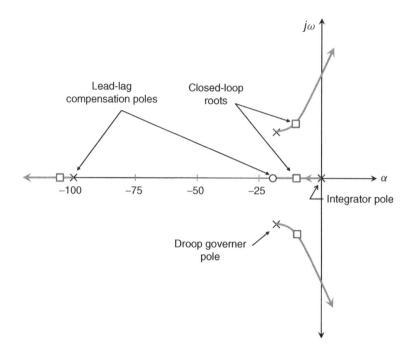

Figure 3.24   Root locus of N-dot system speed control.

There are some issues that should be mentioned regarding N-dot control that should be considered by the fuel-control system designer. Should the engine fail to respond to the slave datum, the inner-loop droop governor will continue to increase fuel flow thus aggravating the problem and driving the compressor into stall. It is therefore common practice to include some form of stall detection and recovery algorithm as part of the control system.

During the engine start phase, it may also be appropriate to use an alternative control methodology until the engine has reached steady-state idle speed operation.

The issue of dealing with a hot re-slams or throttle bursts (which lead to the consideration of N-dot acceleration techniques) can be dealt with using a compensator based on a simple model of the physics involved.

The process of rapidly reducing the power level of a gas turbine from a high-power setting creates a thermal imbalance in both the compressor and the turbine. It has been demonstrated [3, 4] that not only does the steady-state operating line move toward surge due to a re-matching of the compressor and turbine, but the surge line is also reduced by local re-matching of the individual stages of the compressor.

This phenomenon is bound up in the changes in blade tip clearances and in the continuous exchange of energy between the incoming air and the metal of the engine. A useful mechanical analogy of this process is shown in Figure 3.25.

Making the mental leap that energy has mass, the relationship between compressor surge and steady state is influenced by the energy imbalance in the thermal reservoir. The rotor speed, which is proportional to the thermal energy content of the engine metal, is regarded as an input to the reservoir. The speed demand from the controller is indicated

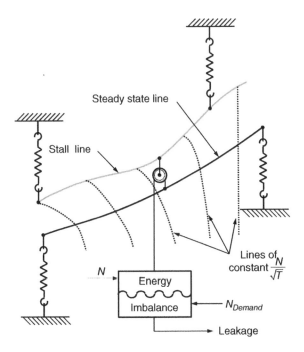

Figure 3.25 Mechanical analog of the 'thermal reservoir' concept.

as an energy output. In addition, there is a leakage flow due to the heat transfer between casings and the surrounding air, both outside the engine and flowing through it.

If we consider a sudden reduction in speed demand, the inflow will remain higher due to the finite time required to decelerate the engine and so the reservoir will fill up. This excess of energy will pull the compressor surge line down and the steady-state running line up; true-steady state conditions will only be achieved by thermal leakage, which is considerably slower than the transient itself. As a consequence, the potential to accelerate the engine during this period will be reduced.

By considering the reservoir as a control volume, we can define the net rate flow of energy into the reservoir as follows:

$$\frac{dE}{dt} = k_p \left(N - N_d \right) - k_\lambda E \tag{3.21}$$

where E is the instantaneous energy imbalance in the reservoir. In this formulation only positive values of $N - N_d$ are considered. By relating this to the reduction in the acceleration potential, it becomes possible to modify a scheduled fuel acceleration limit as:

$$W_{\text{FD}} = W_{\text{F}} - E. \tag{3.22}$$

Since E will always reduce to zero in thermal equilibrium, this form of compensator will only apply during hot re-slam conditions. A block diagram representation of this compensator is shown in Figure 3.26.

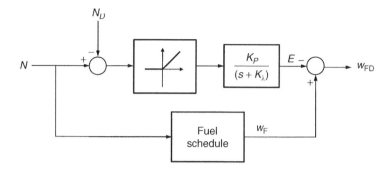

Figure 3.26   Hot re-slam compensator using heat soakage.

3.2.4   Compressor Geometry Control

As discussed in Chapter 2, modern axial compressor designs often depend upon the control of IGVs and/or VSV control of one or more of the lower pressure stages to ensure surge-free operation of the compressor during acceleration of the engine. These devices are typically scheduled as a function of engine non-dimensional parameters, such as corrected speed or compressor pressure ratio. Since one or more of these parameters are often available as part of the acceleration and deceleration schedule control function, these same parameters can be readily adapted to generate variable geometry vane position schedules that can be used as command signals to a vane servo actuator (as depicted schematically in Figure 3.27).

Since these variable vanes are located around the complete circumference of the compressor, a novel actuation method is usually used. This involves an actuation ring around the engine case that connects to individual drive links for each stator vane, as indicated in the conceptual drawing of Figure 3.28.

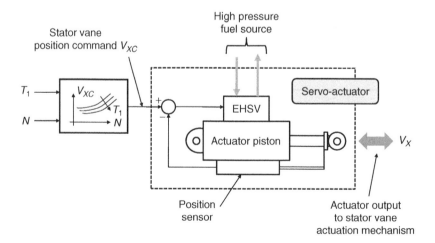

Figure 3.27   Variable stator vane control schematic.

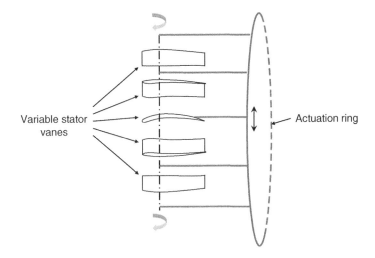

Figure 3.28   Variable vane actuation mechanism concept.

3.2.5   Turbine Gas Temperature Limiting

An important feature of gas turbine operation in aircraft propulsion applications is the fact that they operate for long periods close to their mechanical and aerothermodynamic performance limits. This is driven by the need for operational efficiency, which requires the highest possible gas temperatures for the thermodynamic cycle. This in turn becomes a major engineering design challenge with regard to the materials and cooling techniques used in turbine design and construction.

The implications of this aspect of aircraft gas turbine application is that the life of the hot section of the engine becomes a major operational driver, since it is closely related to engine maintenance costs and hence the overall cost of ownership.

As a result, great attention is paid to the need for the control of hot-section operational limits either via the flight crew constantly monitoring TGT or by dependence upon some form of automatic TGT limiting function.

TGT limiting is employed to provide the engine protection during both power transients and during steady-state operation. During engine acceleration, one common technique used to provide TGT limiting is to modify the acceleration schedule in the high-power operating regime to follow lines of constant TGT. This is illustrated in Figure 3.29 (a repeat of Figure 3.12 but with the acceleration schedule modified at high gas generator speeds). Note how the TGT limit line takes over before the governor droop line, thus eliminating the sharp peak in fuel flow that occurs without this TGT-limiting feature.

With the introduction of electronic engine controls, more direct methods of TGT limiting have become the norm. Some of the issues and design concepts that have been adopted are described in the following text.

While the objective is to limit the high pressure turbine inlet temperature (TIT), which is the high point of the thermodynamic cycle, this is difficult to measure directly. An estimate of this parameter is therefore usually inferred using thermocouples that measure gas temperature at some lower temperature stage, for example, low pressure TIT or exhaust gas temperature (EGT).

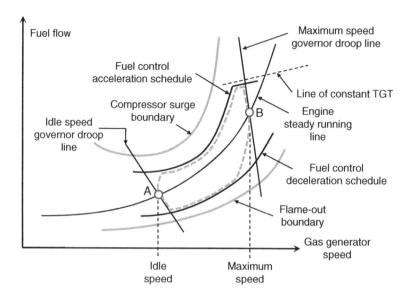

Figure 3.29   Acceleration limiting with TGT limit scheduling.

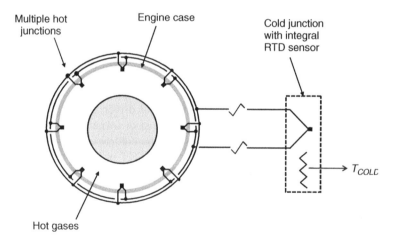

Figure 3.30   TGT thermocouple arrangement.

Figure 3.30 shows how thermocouples are used in the measurement of TGT and/or EGT in gas turbine applications. Typically, several hot junctions are situated around the periphery of the hot gas stream. The cold junction is located in the instrument or control unit. Since the current flowing between the hot and cold junctions is proportional to the temperature difference between the two junctions, it is necessary to measure the cold junction temperature so that the hot junction temperature can be determined accurately.

A problem associated with the use of thermocouples to measure TGT in a closed-loop control system is the fact that there is a significant time delay involved in the sensing process. The more rugged the thermocouple design, the slower it responds to changes in temperature. In addition, the time constant characterizing the thermocouple varies with operating conditions. This makes it difficult to compensate for in any reliable or consistent fashion.

To circumvent this problem, some advanced military engine applications use optical pyrometer technology to measure turbine blade temperature directly from the infra-red radiation emitted by the turbine blades; however, this high-technology approach is expensive and is not in common use in most aircraft gas turbine applications.

3.2.6 Overspeed Limiting

The most critical failure mode associated with the gas turbine in aircraft operation is the rotor burst failure that can occur if one or more of the compressor-turbine spools are allowed to exceed their mechanical design limits.

While every effort is made in aircraft design to structurally contain such an event within the engine nacelle, the design assumptions for aircraft equipment in the potential path of rotor burst debris must make the assumption that such debris has infinite energy. As part of the aircraft design process, provision must therefore be made to ensure continued functionality of all critical systems within the aircraft should such an event occur.

On the prevention side, the gas turbine fuel system is required to provide overspeed protection by cutting off fuel flow to the engine, should an overspeed situation develop.

The functional integrity design goal for this function is that the probability of occurrence of an engine overspeed is less than one in a billion flight hours. This is accomplished by having an overspeed detection and fuel cut-off function that is functionally independent of the primary speed governing system. In some applications, a separate topping governor is provided to improve functional integrity.

In any case, any overspeed system solution should ensure that dormant failures cannot occur and remain undetected, thus reducing the integrity of the overspeed function.

3.3 Fuel System Design and Implementation

This section addresses the equipment design and implementation issues associated with gas turbine fuel systems, citing specific examples.

The functional aspects of a typical fuel control can, in most cases, be considered as comprising two basic sections:

- the control and computation section; and
- the fuel metering section.

Prior to the development of the transistor which led to large-scale integration of transistors and eventually single-chip microprocessors, fuel control technology was confined to the area of hydromechanics. This led to the development of complex, compact servomechanisms achieving remarkable operational reliability considering the harsh environment associated with the aircraft gas turbine. There was a brief but unsuccessful attempt to apply electronics in gas turbine control in the 1950s. The operational environment proved to be too demanding for the vacuum-tube electronics technology of the day.

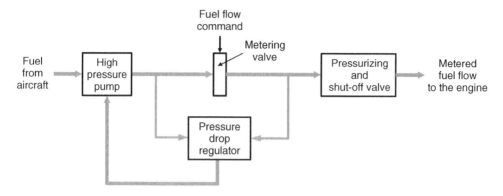

Figure 3.31   Fuel metering unit conceptual diagram.

The fuel metering section of the typical fuel control has evolved into a consistent design approach, summarized by the high-level conceptual diagram of Figure 3.31.

The figure shows the most common design approach in service today. The HP pump is an engine-driven positive displacement gear pump whose output flow is proportional to drive speed. To control fuel flow to the engine, a pressure drop regulator maintains a constant pressure drop across the fuel metering valve by spilling excess flow back to the pump inlet. Metering valve position, and hence flow area, is therefore proportional to the commanded fuel flow. Finally, a pressurizing and shut-off valve isolates fuel from the combustor when the engine is shutdown and prevents fuel flow to the engine until sufficient pressure is available during the start-up phase.

Alternative architectures using variable displacement pumps have been used in the past; however, they are more sensitive to contamination than the gear pump which is preferred today.

A more detailed description of the fuel pumping and metering section of the fuel control is presented in Section 3.7.2.

We continue with a historical overview describing how the technologies associated with fuel control systems have changed, in particular with the advances in electronics that have occurred over the past 50 years.

In the review of system architectures which follow, the evolution of fuel system solution is presented to provide the reader with some appreciation of the design challenges involved. Note that the costs of a control solution do not scale well with size, making small engine control designs quite different to those for a larger engine. The reasoning behind the past several generations' fuel system designs, brought about by the rapidly changing technologies as they became available, is also explained.

Fuel system solutions for a number of system design concepts are presented, covering the control, pumping, and metering aspects of fuel systems.

Finally, certification issues associated with gas turbine fuel systems in aircraft service are discussed.

3.3.1 A Historical Review of Fuel Control Technologies

The implementation of gas turbine engine fuel control systems has changed dramatically over the past half-century.

Early control designs used mechanical interconnections and mechanisms to provide the engine fuel control function, because of the criticality of the function itself. Early attempts during the 1950s to use the obvious advantages of electronics to provide some of the more complex fuel control and management functions were fraught with reliability problems. These were due to the inability of electronic components of the day (such as thermionic valves/vacuum tubes) to withstand the hostile environment involved.

This temporary diversion brought with it some exceptional hydromechanical technologies including 'fluidics', which were envisioned at the time as the answer to the environmental reliability problems of electronic controls.

Hydromechanical control computing continued to be the primary technology employed by the engine fuel control community through the 1970s, driven by their operational success in terms of the impressive functional reliability and availability that was demonstrated in both commercial and military applications for several decades. These hydromechanical controls employed many innovative concepts that utilized flyweight speed sensors driven by the reduction gear connected to the HP rotor, 'nutcracker' servos to linearize outputs from non-linear parameters, three-dimensional cams, and multiplying mechanisms to provide the necessary computations associated with the development of acceleration/deceleration schedules and speed governing.

These devices used HP fuel as the motive pressure source. Figure 3.32 depicts examples of this approach in simplified schematic form.

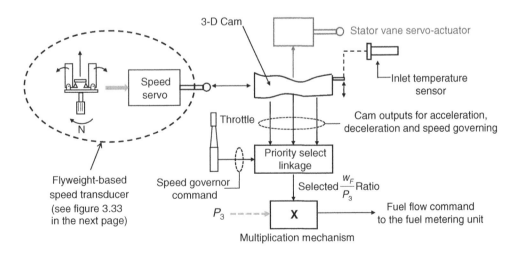

Figure 3.32 Typical hydromechanical computation mechanisms.

In the figure, the 3D cam is modulated in one dimension by the speed servo, which converts the speed-squared force signal from the flyweight sensor to a linear displacement and, in the other dimension, by an engine inlet air temperature servo.

The cam surfaces generate fuel flow ratio contours that become inputs via cam followers to linkages that provide the required priority selection described previously. The selected fuel flow ratio output goes to a multiplication mechanism, from which the fuel flow command is generated. The same 3D cam can also be used to generate a VSV position schedule in the form of a cam surface, which is a function of speed and inlet temperature, should this be required.

Figure 3.33 is a more detailed depiction of the speed transducer servo-mechanism. As shown, the flyweight device generates a speed-squared force on the spool valve which ports HP fuel to a servo piston. The nutcracker feedback arrangement varies the spring feedback stiffness seen by the spool valve as a function of the servo output position. The result is a linear relationship between speed and piston displacement.

There are also many more innovative fluid-mechanical fuel control concepts that have been developed over the past several decades, one of which is worth mentioning here. The challenge for small engines (say, less than 1000 HP) is that the cost of the fuel control system is typically limited to a percentage of the total engine cost; in most cases, the complexity of the 3D cam and complex mechanism approach may be unacceptably high. The only way to address this problem is to consider compromise solutions to the control system problem using innovative design techniques.

One such example is the DP-F2 control developed by the Bendix Corporation (now Honeywell) in the late 1950s for the PT-6 engine. This control concept uses compressor discharge pressure which flowed through the control unit through various orifices to develop a simple w_F/P_3 schedule. A schematic of the control section of the DP-F2 fuel control is shown in Figure 3.34.

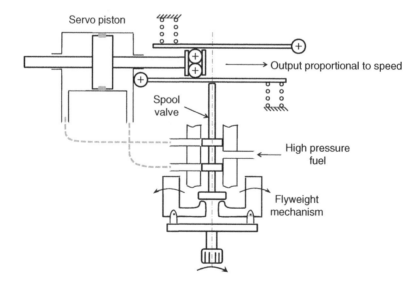

Figure 3.33   Speed transducer servomechanism schematic.

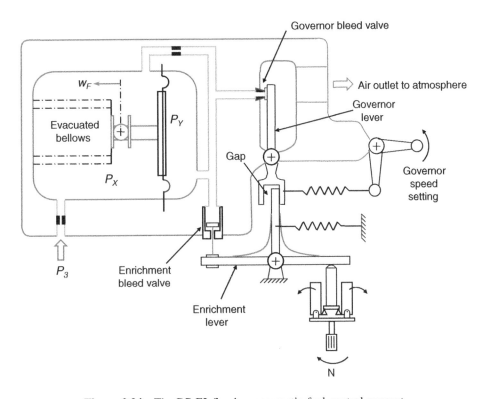

Figure 3.34 The DP-F2 flowing pneumatic fuel control concept.

The best way to understand this simple system is to initially ignore the enrichment lever and its bleed valve assembly. In this arrangement, with the governor valve forced into the closed position by the set speed lever tension spring, both P_X and P_Y pressure become equal to the compressor discharge pressure P_3. The difference in effective diaphragm area afforded by the existence of the evacuated bellows results in the fuel flow command point being directly proportional to P_3. The acceleration fuel flow therefore follows the control mode $w_F/P_3 = $ constant.

As the engine speed increases and the flyweight speed sensor force becomes large enough to overcome the spring preload, the governor bleed opens reducing P_Y pressure until the engine speed equals the set speed command.

This simple control law is, in practice, a bit too simple. By setting the constant so that the compressor surge boundary is not penetrated under all conditions, the acceleration rate at the higher gas generator speeds becomes much too sluggish. The answer is to provide a two-step acceleration schedule with a lower-level setting to circumnavigate the surge boundary followed by a higher-level setting to increase acceleration rates beyond the surge bucket. In the DP-F2 control, this is accomplished using the enrichment system. Here an enrichment lever is included between the flyweight sensor and the governor lever with its own tension spring.

When the speed sensor develops sufficient force to overcome the enrichment lever preload, the enrichment lever moves through the gap between the two levers. In doing so it closes off the enrichment bleed, thus raising the acceleration schedule for the higher engine speed range. As the speed approaches the set speed, the governor bleed valve opens as before to provide speed governing at the commanded set speed.

This example shows that simple, innovative approaches are possible if some compromise from the ideal solution can be demonstrated as acceptable in operation. In the above example, the control 'computer' comprises a simple aluminum housing plus a couple of metal bellows and a few pneumatic bleed orifices.

An important aspect of this design is that the interface between the wet (fuel) and dry (pneumatic) sections of the fuel control unit uses a torque tube, thus eliminating the need for a dynamic seal between the two sections. This eliminates the potentially unsafe failure modes which could result in hot compressor air getting into the fuel section of the control.

The advent of the transistor and its miniaturization led to very large scale integration (VLSI) technology and eventually single-chip microprocessors. This, in turn, led to a revolution in the involvement of electronics in engine controls as well as throughout the aerospace industry that continues to this day.

The ultimate accomplishment in gas turbine control was the introduction of the full-authority digital electronic control (FADEC), the equivalent of 'fly-by-wire' in aircraft flight control systems.

In the FADEC-based control, the computations provided by all of the complex mechanisms, 3D cams, etc. can be implemented in software. With this powerful new approach, there is almost no limit to the complexity of the control algorithms that can be implemented.

Today, almost all modern gas turbine propulsion systems employ FADECs for the fuel control computation function. To give some perspective to this commentary, Figure 3.35 shows some of the major technology milestones regarding the application of electronics in engine fuel control systems. As shown, electronic controls evolved from limited authority trimming devices in the 1960s to full-authority controls in the 1970s. Military engines were the first to employ full-authority electronic controls. One notable example is the F-100

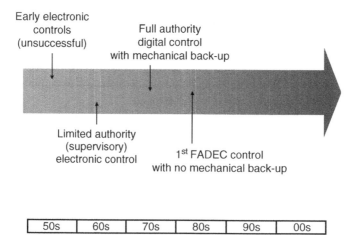

Figure 3.35   History of engine control technology.

engine which uses the digital electronic engine control (DEEC) to provide full-authority control of the gas generator and the afterburner. The DEEC is a simplex (single channel) microprocessor-based control with is a mechanical back-up system to accommodate any major failures of the electronic control system and to provide the pilot with continued control of the engine (albeit in a degraded mode).

In the commercial world, the application of supervisory electronic control became a common technique whereby the basic hydromechanical fuel control system could be trimmed within a limited authority to provide engine exceedance protection including:

- TGT limiting;
- torque limiting in turboprop applications; and
- combustor pressure limiting (during cold day operation).

These control features were aimed at minimizing the maintenance penalties associated with operational oversight where the crew may, on occasion, fail to manage engine power settings appropriately. Supervisory electronic controls were therefore aimed primarily at crew workload reduction and reducing maintenance costs.

One problem with this situation is the added acquisition cost, since the most common design approach was to maintain the basic hydromechanical control with its 3D cam, flyweight governor, and complex servomechanisms and to add an electronic control unit that would trim the pilot's throttle setting via some effector (typically a stepping motor or a proportional solenoid).

For small engines, the cost budget for the fuel control system as a percentage of the overall engine cost was a challenge to the control system supplier. This was perhaps the main reason for the success and popularity of the DP-F2 flowing pneumatic control described above.

One innovative solution developed by Hamilton Standard (now Hamilton Sundstrand) in the late 1970s, following the introduction of the first single-chip microprocessor by Intel Corporation, was a fuel control system aimed at the small engine market. This comprised a simplex (single channel) electronic control unit which trimmed a simple hydromechanical FMU, providing full-authority control during normal operation but having a simple mechanical back-up capability in the event of a failure of the electronics.

A description of this design approach is worthy of inclusion here. A conceptual schematic of this small engine fuel control is shown in Figure 3.36. (The definition of a small engine in this context is less than 5000 lb thrust or 1500 SHP). The functional concept for this control system is to provide a minimum fuel flow schedule to the engine determined by a combination of throttle lever position (PLA) and compressor discharge pressure P_3 which can be 'trimmed upward' by the electronic control. This is accomplished by having the electronic control unit drive a torque motor which adjusts the pressure drop regulator setting by modulation of a variable restrictor in the pressure drop regulator control line.

When the electronic control input to the torque motor is zero, the controlled flow restrictor is closed off and the pressure drop regulator provides a fixed, predetermined ΔP across the fuel metering valves. When the electronic control unit is in control, commands to the torque motor vary the restrictor area. This modulates the pressure drop regulator set point upwards and increases fuel flow to the engine.

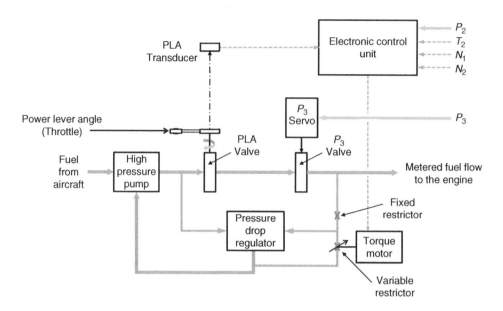

Figure 3.36 Small engine digital electronic control with manual back-up.

Acceleration and deceleration limiting is provided by the electronic control using the N-dot slave datum technique described in Section 3.2.4 above. In this case, the prevailing N-dot limit is a function of gas generator speed and inlet air temperature and pressure.

Isochronous speed governing is also provided by the electronic control with gas generator speed proportional to throttle position. In the turbofan application, this same control can provide isochronous fan speed governing as well as speed synchronization between engine pairs.

In the manual back-up mode, the torque motor current is set to zero and the pressure drop regulator re-sets to its base fixed value. As the throttle valve (PLA) is increased, fuel flow increases but the maximum value is constrained by the prevailing value of P_3. If PLA is set to its maximum value, the fuel flow to the engine is therefore dominated by the position of the P_3 servo-driven valve such that:

$$w_F = KP_3. \tag{3.23}$$

An over-travel of the throttle is provided to allow the pilot to obtain full power when in manual back-up mode.

It was not until the early 1980s that sufficient reliability had been demonstrated to abandon the mechanical back-up and launch the true FADEC era.

The following section provides a detailed coverage of the fuel pumping and metering section of the gas generator fuel control system and its interface with the modern FADEC.

3.3.2 Fuel Pumping and Metering Systems

As mentioned at the beginning of this chapter, high fuel pressures are necessary to provide effective fuel nozzle atomization performance for combustion. These pressures may be well in excess of 1000 psi above the prevailing inlet total pressure.

For this reason, positive displacement HP fuel pumps are typically employed as the primary fuel pumping element for delivering fuel to the gas generator combustor.

There are three major pumping technologies that have been applied to the HP fuel pump requirements for the gas turbine propulsion systems, specifically:

- the variable displacement swash-plate piston pump;
- the variable displacement vane pump; and
- the fixed displacement gear pump.

The first of the above pumping concepts was employed extensively in some of the earlier gas turbine fuel system designs; however, this type of pump proved to be significantly intolerant of contamination levels typical of aircraft fuel systems and, as a result, poor in-service reliability drove the need for alternative solutions.

The variable displacement vane pump offered an alternative to the piston pump concept; however, the vane pump failure mode is not attractive since a vane failure can mean instant loss of HP fuel and an engine shutdown situation.

The fixed displacement gear pump option has become the preferred pump technology over the past 50 years, due to its ability to tolerate significant levels of contamination while demonstrating impressive in-service reliability. Failure modes with this type of pump are primarily manifested by pressure performance degradation and therefore easily tolerated from an in-service operation perspective.

Figure 3.37 shows a top-level schematic of a typical gas generator fuel system for a FADEC-controlled engine showing the HP gear pump concept, the FMU, and an associated VSV actuator.

Also shown in the figure are some of the ancillary functions that are typically included in the fuel handling section of the gas generator, including:

- fuel filtration to protect both the fuel combustion nozzles as well as the fuel metering and actuation servo-mechanisms;
- cooling of the engine bearing and auxiliary gearbox lubrication oil; and
- draining of the fuel combustion manifold after engine shutdown to prevent nozzle coking.

The HP fuel pump section comprises a centrifugal backing pump which provides fuel to the gear stage after passing through a fuel oil cooler heat exchanger and LP filter. This filter typically has a bypass valve and an impending bypass indicator or delta-P switch which may be monitored by the FADEC. This is not shown in the figure for clarity. The backing pump receives fuel from the aircraft within specified limits regarding pressure relative to the prevailing vapor pressure of the fuel, vapor/liquid ratio, and viscosity. The pressure rise across this stage is typically of the order 100 psi which ensures that cavitation at the input to the gear stage does not occur.

The HP gear pump has an integral relief valve to prevent over-pressurization of the FMU.

The FMU meters the correct amount of fuel to the engine, spilling the excess flow back to the inlet of the fuel-oil cooler. The FMU also provides a HP fuel source for any servo-actuators required by the system. Metered fuel passes through a mass flowmeter and HP filter on its way to the fuel manifold and combustion nozzles. After engine

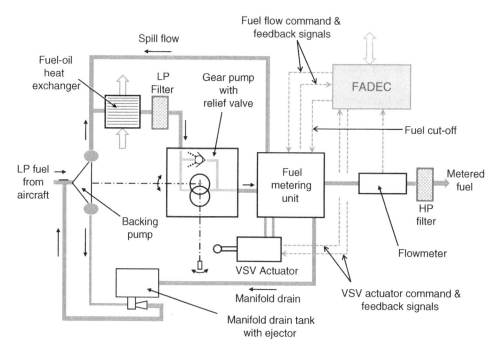

Figure 3.37 Gas generator fuel pumping and metering system overview.

shutdown, the FMU allows the fuel manifold to drain into an ecology tank where it is transferred to the backing pump inlet following engine restart using an ejector pump fed by interstage pressure.

In the example shown, a FADEC provides fuel flow command and fuel cut-off signals to the FMU. The VSV actuator is also controlled by the FADEC. The flowmeter is used to provide fuel flow information to the avionics and flight deck and is typically not used in the engine fuel control function. In the example shown, the flowmeter interfaces with the FADEC which transmits fuel flow and other fuel system status information to the aircraft via a digital data bus.

It is worth noting here that there is a fundamental problem with the gear pump-based fuel metering system, illustrated by the pump performance graph of Figure 3.38. As shown, the gear pump delivers flow in proportion to the pump's drive speed which comes from the engine HP shaft through a reduction gearbox.

As delivery pressure increases, leakage across the gear faces results in a reduced flow for any given speed; however, the basic flow/speed slope remains essentially the same. If we superimpose typical engine performance curves onto the pump performance graph, we can see that the engine start flow requirement determines the pump size. As a result there is a substantial excess flow at the higher engine speeds which must be spilled back to the pump inlet. This spill flow is a major generator of heat and is very high at the altitude cruise condition which combines high pump drive speeds with relatively low fuel flow requirements. In commercial transport aircraft, this represents more than 90% of the system operational time.

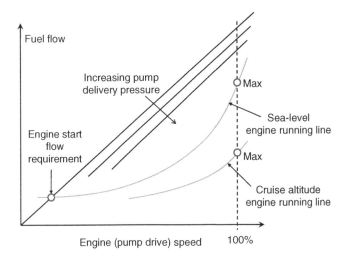

Figure 3.38 Typical high-pressure gear pump performance.

Today, a number of new, rugged, variable displacement HP fuel pumps as well as fuel metering pump concepts are being evaluated in an attempt to eliminate this power loss and heat dissipation problem.

A more detailed schematic of a typical FMU is shown is Figure 3.39. Referring to the figure, HP fuel first passes through a wash-flow filter which provides clean HP fuel to any servo-actuators in the control system.

The metering valve shown is a simple spool valve whose position is controlled by a torque motor using the HP fuel source. A linear variable differential transformer (LVDT) provides metering valve position information back to the FADEC.

By maintaining the pressure drop across the fuel metering valve constant, the flow area of the metering valve will be directly proportional to volume flow through the valve. This is accomplished by a pressure drop regulator which spills excess flow back to the gear pump inlet (via the fuel-oil cooler). Since it is desirable to control mass flow rather than volume flow to the engine, it is common practice to use bimetallic washers behind the pressure drop regulator spring to compensate for fuel density variations with fuel temperature, thus providing effective mass flow metering. The pressure drop regulator can also be used to compensate for different fuel types by modifying the pressure drop setting via an external adjustment.

The fuel output from the metering valve passes through a shut-off and pressurizing valve which remains closed during starting until a certain minimum fuel pressure has built up. During normal operation, this valve remains fully open. If the throttles are closed to the fuel cut-off position, the fuel cut-off solenoid is energized and selects HP fuel to the right side of the pressurizing valve. This forces it to close and shuts off fuel to the engine. The manifold drain valve can now open, allowing the combustor manifold to drain into the ecology tank.

The overspeed trip system may use the same hydromechanical equipment, however any sensing or electrical signaling must be sufficiently redundant for functional integrity reasons.

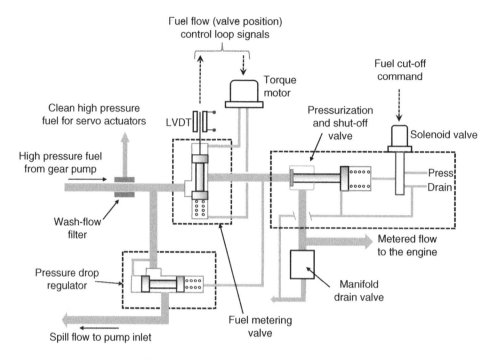

Figure 3.39   Fuel metering unit schematic.

After leaving the FMU, metered fuel flow is fed to one or more fuel manifolds which are connected to as many as 20–30 nozzles inside the combustor.

The design and implementation of the combustor/nozzle arrangement is highly specialized and beyond the scope of this book. Sophisticated design tools are used, including computer fluid dynamics (CFD) to study airflow characteristics, nozzle spray patterns, fuel droplet sizes, etc. over the complete engine operating range to ensure optimum mixing of fuel and air. The goal is to achieve complete combustion without hot spots that can lead to the generation of oxides of nitrogen.

To accommodate the wide fuel flow range from starting to maximum power, many nozzle designs have two outlets: a starter outlet and a main outlet. Figure 3.40 shows the start/main flow characteristics indicating how flow is switched from start to main as a predetermined nozzle pressure drop is met. This type of fuel nozzle is referred to as a 'duplex nozzle'. An alternative to having dual outlet fuel nozzles is to have separate starting nozzles fed from a separate start flow manifold.

The technology associated with combustor/nozzle designs continues to escalate in order to address the stringent engine emission requirements that become more demanding with each new generation of engine.

An ignitor system is used to light off the engine during starting. The ignitors are also selected when flying through heavy rain where there is a possibility of a flame-out occurring.

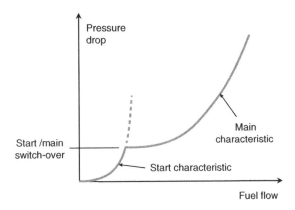

Figure 3.40   Fuel nozzle characteristics.

3.4   The Concept of Error Budgets in Control Design

The most fundamental notion of control requires the measurement of engine parameters (typically rotor speeds, temperatures, and pressures). In a typical droop-speed governor, the measured speed is compared to a desired value. The difference is used, together with inputs from other measured engine parameters, to determine a command to a fuel value which will change the input (fuel flow) to the system in order to move the measured speed toward a favorable comparison with the desired value. A block diagram of a simple droop governor typical of many of the traditional hydromechanical fuel controls is shown is Figure 3.41.

Consider the primary objective before examining Figure 3.41 which, in this case, is the control of engine speed to within an acceptable accuracy. Figure 3.42 is a plot of the steady-state fuel flow for a typical engine as a function of rotor speed. This figure indicates that the slope of the curve changes with rotor speed; however, at maximum speed (100% N) a 1% change in rotor speed requires a 4% change in fuel flow. This does not sound too difficult, until the rate of change of TIT with speed is examined. A calculation of this parameter indicates that a 1% change in rotor speed is accompanied by a 50 °F change in TIT. By standards of engine reliability, this is a very large change indeed; it is therefore desirable to control rotor speed within 0.1% and preferably 0.05%.

Translating this requirement to the individual components shown in Figure 3.41 suggests that the inaccuracy contribution from all components must small enough such that the overall accuracy of 0.1% (or less) control of rotor speed is achieved. The starting point must therefore be the level of accuracy with which we can measure rotor speed. While many tolerance-related errors can be adjusted out during ground calibration testing, variations in inlet conditions will require some compensation of the speed governor in order to maintain the required relationship between PLA and engine speed. This factor brings into play the accuracy of the temperature and pressure sensors.

The older hydromechanical controls required that all of the sensed parameters and control functions were represented by mechanical analogs such as cams, springs, and servo-mechanisms which recognize friction, inertia, and all other aspects of such

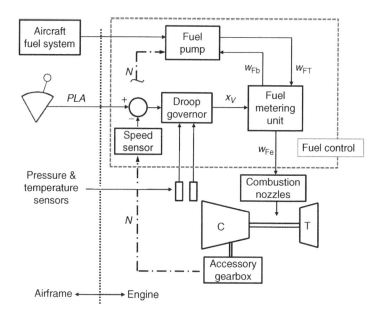

Figure 3.41   Basic engine control functions.

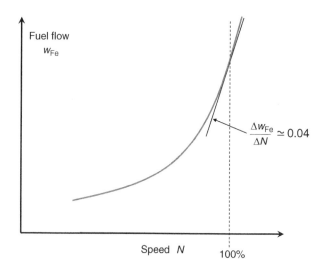

Figure 3.42   Typical engine fuel flow versus rotor speed.

mechanisms which can and do affect accuracy. It is remarkable how well these hydromechanical controls were able to support the requirements of the aircraft gas turbine in terms of performance and reliability over the past 60 years or so. With the introduction of the modern FADEC, some of the above-described accuracy problems are eliminated. Speed governors can easily be made isochronous with three-term controller algorithms incorporated to ensure good response and stability.

The output of the droop governor becomes a position demand to a fuel metering valve whose position is directly proportional to fuel flow. This is accomplished as described previously by a pressure drop regulator, which holds the pressure drop across the fuel metering valve constant and bypasses any excess fuel flow back to the inlet of the fuel pump. The fuel metering section of the fuel control therefore has its own issues of frequency response and accuracy.

Finally, the fuel flow is delivered to the fuel manifold and is metered to the engine through fuel nozzles, thus determining the speed of the engine. The engine drives the accessory gearbox upon which the fuel control is typically mounted, providing mechanical speed inputs to the fuel pump and droop-governor mechanism.

At some point in the design process, a selection must be made of the major components that will be used to implement the design. During this selection process, due attention must be paid to the accuracy of the individual components. From this basic data, it would be advisable to develop what is commonly referred to as an error budget so that:

1. system calibrations can be made; and
2. the overall accuracy of the system can be determined.

Some discussion of system errors and their sources is in order to support this effort.

3.4.1 Measurement Uncertainty

The process of measurement involves some type of probe which will stimulate a specific feature of the item being measured and from that stimulus, a measurement can be obtained. Alternatively, a measurement is possible by receipt of a signal which is emanating from the object, independent of whether a measurement is required or not. A probe measuring the total pressure of a flow is an example of the former, while a receipt of infrared radiation from which a temperature can be deduced is an example of the latter.

In either case, it must be assumed that the measurement obtained has some error. A portion of this error is fixed (commonly referred to as bias error) and a portion is random (commonly referred to as precision error). The bias error indicates a known and predictable deviation from the true value as shown in Figure 3.43. As long as the relationship is known, the difference can be accounted for through some form of calibration of the device and, if handled properly, has no particular effect on overall system accuracy.

The second component of measurement accuracy is random in nature and must be recognized as such in any design. The normal practice is to assume a normal Gaussian distribution in this component of the measurement as shown in Figure 3.44. One standard deviation is the usual measure of the precision error. This is approximated in test as:

$$\sigma_S \equiv \sqrt{\frac{\sum_{i=1}^{n}(x_i - \overline{x})^2}{n-1}} \tag{3.24}$$

where x_i is an individual measurement, $\overline{x}$ is the average of all measurements, and n is the total number of measurements.

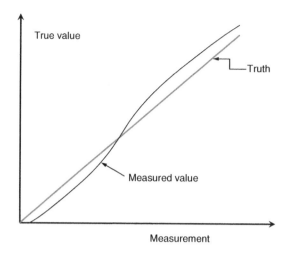

Figure 3.43   Typical bias error.

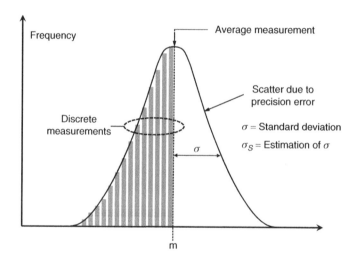

Figure 3.44   The random nature of precision errors.

The larger the variation in the measurement value, the greater the uncertainty in any single measurement. This is seen as discrete errors in a digitized system, whereas it is seen as noise in an analog system.

3.4.2   Sources of Error

There are a significant number of sources of measurement error that should be considered during a design/component selection process. These will be discussed briefly in the following sections.

3.4.2.1   Geometric Errors

As a general rule, the control system designer strives to minimize the number of sensors fitted to an engine. With this in mind, the assumption of one-dimensional flow is employed wherever possible.

Such a design condition carries with it the practical reality that there may be variations across a flow that are ignored. For example, a single pressure measurement made of engine inlet conditions must ignore flow distortions that may occur across the inlet which may further change in severity depending on the flight regime. A similar condition prevails in industrial and marine installations.

The type of error that is associated with the assumption of one-dimensional flow is generally of a bias type, which therefore can be calibrated out of the system provided the bias error can be known over the entire operating regime. This is not always the case, and considerable degradation in overall system performance can occur where the one-dimensional assumption has not been thoroughly examined. This is particularly true where the bias varies with operating conditions.

A major exception to the one-dimensional assumption is the measurement of EGT. This is especially true for single-shaft turbo props which operate at a constant speed and control power through the control of EGT. In this case, a harness is fitted with as many as 16–32 thermocouples thus obtaining an average temperature around the circumference of the annular exhaust area (see the example in Figure 3.30 above).

3.4.2.2   Transient Errors

A significant source of error of particular concern to the controls designer is that of transient or time-dependent errors. Figure 3.45 is a characterization of the problem. In

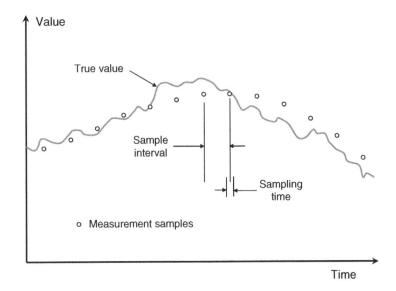

Figure 3.45   Examples of transient measurement error.

this diagram there are several phenomena of interest. The so-called true value of the measurement has a small but noticeable fluctuation which is apparently inherent to the physical phenomenon. For example, this could be associated with high-frequency pressure fluctuations at the compressor delivery. At the same time, there is a major transient excursion underway which is the primary measurement of interest.

In such a circumstance, the high-frequency fluctuation is rightfully regarded as noise which must be filtered out of the final signal to be processed by the control.

The measurement indicated in the figure is being obtained at a fixed sample rate, and it is apparent that the level of filtering imposed by the overall measurement technique has additional lags such that it is unable to follow the large-scale transient shown.

This generalized presentation highlights the need for sufficient signal conditioning to cope with the system noise; however, it must still span the range of frequencies that will allow acceptable accuracy of control both transiently and in steady-state governing.

3.4.2.3 Sampling Errors

Modern controls are dominated by the digital microprocessor which converts all signals to a digital format. The digital processor is a serial device and this feature introduces errors peculiar to this process. Conversion of a signal to a digital representation is achieved through an analog-to-digital converter. This device contributes to the total error in several ways.

The conversion accuracy of an analog-to-digital (A/D) converter is specified by the number of binary bits to which it is able to resolve the analog signal. This is defined mathematically as $1/2^n$, where n defines the resolution of the A/D converter in terms of bits. Table 3.1 shows how the inherent inaccuracy of an A/D converter varies with bit-count.

Since a single controller is typically equipped with a single A/D converter, it must poll each sensor separately. This introduces phase errors as shown in Figure 3.46.

The figure shows three sensors (S_1, S_2, and S_3) being polled in sequence. At the start of the process, the A/D converter clamps S_1 and completes its conversion. In the meantime, the physical parameters represented by S_1, S_2, and S_3 continue to move with time. S_2 is converted in sequence and then S_3 is converted. The total elapsed time is ΔT which is divided into three equal parts. As the figure suggests, there is a slewing of the signals with time which must be ignored in the calculation contained within the overall control loop.

A third possible source of error in the digital environment is termed 'aliasing'. The Shannon theory related to aliasing states that the sampling rate must be $2\pi f$, where f is the highest frequency contained in the signal of interest.

Table 3.1   A/D conversion errors.

Bit count	Conversion error $\pm\%$
8	0.3906
10	0.2904
12	0.0031
16	0.0015

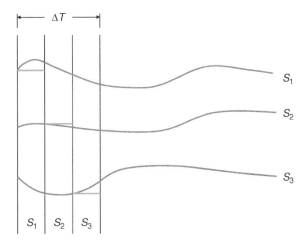

Figure 3.46 Phase errors in digital sampling of several sensors.

Since it is common practice in engine FADEC technology to sample all sensor signals within a 20 ms window, which represents more than $20f$, there is therefore a significant margin of error in capturing the information.

3.4.2.4 The Error Budget

The previous discussion revealed the fact that there are errors associated with measurements. It can be further stated that every element in a control loop will produce a result which, in a manner exactly analogous to the measurement uncertainty, will contain errors. The designer must therefore recognize the propagation and accumulation of errors around the closed-loop process of measurement, computation, and actuation. This is recognized by Abernethy *et al.* [5] and can be expressed by the following equations:

$$S = \pm\sqrt{s_1^2 + s_2^2 + s_3^2 + \cdots + s_n^2} \qquad (3.25)$$

and

$$B = \pm\sqrt{b_1^2 + b_2^2 + b_3^2 + \cdots + b_n^2} \qquad (3.26)$$

where s is precision (random) error of each contribution, b is bias error of each contribution, S is total precision loop error of the system, and B is total bias error of the system.

Based on these simple expressions, an error budget chart can be prepared for any proposed design as indicated in Table 3.2. As can be seen, a distinction is made between the steady-state error and transient errors. Such a distinction clearly highlights that the accuracy of control, during a transient where a temperature measurement is involved in the control calculations (the accuracy by which fuel is metered to the engine), degrades due to the transient errors associated with temperature measurements. This is especially true

Table 3.2 Control loop error budget.

Component	Error in % of absolute			
	Steady state		Transient	
	b	s	b	s
PLA demand signal	2	0.01	2	0.01
Speed sensor	Neglible	0.01	Neglible	0.01
Pressure sensor	0.01	0.1	0.01	0.1
Temperature sensor	Neglible	0.5	Neglible	5.0
Fuel valve position	1.0	0.1	1.0	0.1
Fuel flow delivered	Neglible	0.01	Neglible	0.01
RMS total:	2.25	0.52	2.25	5.0

for single-shaft turboprop engines where speed is held constant and power is controlled directly on measurement of an EGT.

See also [6, 7] on the subject of error budget and measurement practices.

3.5   Installation, Qualification, and Certification Considerations

3.5.1   Fuel Handling Equipment

The HP fuel pump assembly is installed on the auxiliary gearbox on a drive pad that provides a suitable speed range for the pumping hardware. This is usually geared down considerably from the engine shaft speed and includes a number of power off-takes for many airframe and engine systems (see Chapter 8).

Figure 3.47 shows typical accessory gearbox mounting arrangements. In order to illustrate an important design approach used by many high-bypass turbofan engines, the figure shows an LP spool (shaded) comprising a fan together with LP compressor and turbine stages in addition to the gas generator section of the engine. In each case, a tower shaft transmits shaft power from the HP shaft of the engine to the gearbox.

Fan-mounted accessories see a much more benign temperature environment than the core-mounted equivalent. However, the fuel flow through the pumping and metering components in core-mounted hardware applications will ensure that material operating temperatures can be maintained sufficiently low enough for the use of aluminum housings for the containment of fuel handling equipment.

There are of course exceptions in very-high-speed military aircraft where steel or titanium materials may be necessary to tolerate the worst-case local temperature environment.

The FMU may have its own mounting pad or, alternatively, it may be installed on the HP fuel pump. (Some manufacturers supply integrated fuel pump and metering packages.) In many applications, spool valves within the FMU are driven (i.e. rotated either from the pump drive or via a separate accessory gearbox pad) to minimize valve friction and thereby improve servo-accuracy. This is an important aspect of the fuel metering task because of the 'turn-down ratio' involved. Maximum fuel flow at altitude may be 3–4 times lower than at sea level. In order to have, say, ±1% of point fuel flow accuracy at the altitude

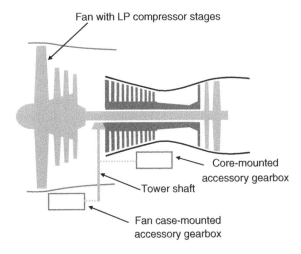

Figure 3.47 Accessory gearbox mounting arrangements.

condition, the device must therefore have a full range accuracy of ±0.25% of full scale; this is extremely demanding when taking into account the harsh operating environment involved. It is typical of today's equipment to provide fuel scheduling accuracies of the order ±2% to ±3% of point over the full range of operation of the engine.

Qualification requirements for fuel pumping and metering equipment are extremely demanding and require demonstration of functional capability under a wide range of operating conditions, such as:

- functional integrity during high levels of vibration;
- high temperature testing including demonstration of structural integrity during a flame test;
- fuel contamination testing; and
- demonstration of pump performance under worst-case fuel inlet conditions, including pressure and temperature extremes and vapor/liquid ratios up to 0.45.

The now defunct military specification MIL-E-5007D is still used by both military and commercial fuel control system sources to establish formal qualification and certification requirements.

The failure modes associated with fuel handling equipment are typically not severe in terms of safety of operation. They tend to manifest themselves as performance degradation over time. Even complex hydromechanical computers using complex cams and fuel-draulic servomechanisms have demonstrated mostly tolerable modes of failure, which are typically limited to schedule shifts and settings falling out of adjustment rather than any loss of critical functionality.

There is however one failure mode that requires fuel system consideration, and that is the mechanical overspeed failure that could lead to an uncontained rotor-burst. This mode of failure is considered catastrophic and, as part of the certification requirement, the probability of this occurrence must be demonstrated to be less than one failure per billion flight hours.

3.5.2   Full-authority Digital Engine Controls (FADEC)

Even with modern solid-state microelectronics, the installation of FADEC controls on aircraft gas turbine engines is extremely challenging because of the hostile environment involved. The primary task of the FADEC structural design is therefore to maintain as benign an environment as possible for the internal electronics in order to provide operational reliability, functional integrity, and safety.

Figure 3.48 shows a conceptual drawing of one approach to FADEC design where the electronics are contained in two physically separate sections of the unit. Pressure sensors are also contained within the unit to take advantage of the relatively benign conditions within. While space for four sensors is shown, three sensors (P_2, P_3, and P_5, for example) may be adequate. In the concept shown, the electronics associated with each control channel are contained on two large circuit boards hard-mounted to the housing in sufficient locations to ensure that there are no low-frequency vibration modes that could be excited by the local environment.

For each channel there are three dedicated electrical connectors which provide sensor signals, data buses, and power access to the unit. A filter section behind the connectors prevents induced currents and electromagnetic interference (EMI) noise from accessing or upsetting the control electronics.

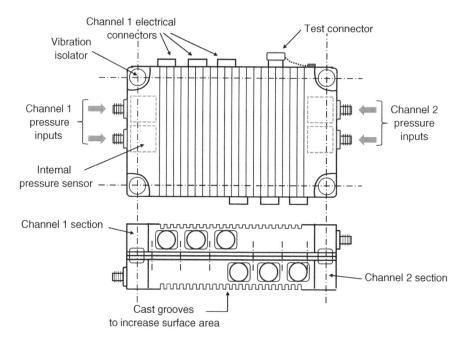

Figure 3.48   Conceptual drawing of a FADEC unit.

A test connector facilitates software downloading and allows the unit to be checked out functionally prior to delivery.

3.5.2.1 Vibration and Temperature

To avoid a degradation in reliability due to vibration, it is usual to mount FADECs on vibration isolators which provide excellent vibration attenuation above about 100 Hz. Care must be taken with the installation design to prevent 'short-circuiting' the isolators through rigid pipe connectors (for pressure sensors mounted inside the FADEC units) or via the wiring harnesses that can be extremely stiff as a result of the metal over-braiding (see next section).

In commercial turbofan engine applications, the FADEC is usually mounted on the fan casing where environmental temperatures are limited to between −40 and +250 °F. During the cruise phase, which represents more than 90% of operational life, the temperature environment is between 0 and +100 °F which is quite benign for electronic equipment. In the example shown, the housing contains a large number of grooves (which should be mounted as near to vertical as possible) to improve the convection cooling efficiency.

Since military engines do not utilize the large fans typical of the commercial turbofan, their FADECs are typically mounted on the core section of the engine. In this case, fuel cooling is necessary to keep operating temperatures within the unit at reasonable levels.

3.5.2.2 Lightning and HIRF

Lightning events can induce large current transients in metal structures and wiring harnesses. Similarly, high-intensity radiated frequencies (HIRFs) from high-power radar transmitters can couple electrical energy into wiring harnesses, thereby accessing the electronic equipment via their connectors with the ability to upset or disable sensitive electronic circuitry. The FADEC environment is particularly challenging since, in today's modern composite material engine nacelles, there is no surrounding aluminum structure to provide the inherent 'Faraday cage' protection. In order to tolerate this environment, FADEC harnesses must be fitted with a metallic over-brade carefully bonded to the connector back-shells at each end of the harness. The FADEC enclosure must also be fabricated so that flange interfaces are designed with good surface finishes and tight bolt spacing to prevent access from very-high-frequency EMI. The enclosure must be bonded to the engine case, taking care not to compromise the vibration isolators.

3.5.2.3 Structural Integrity

The FADEC must meet structural integrity standards of engine-mounted equipment, which includes explosion-proof testing for electronic equipment. Structural integrity must be maintained should an explosion internal to the FADEC occur. This ensures that the effects of such an event cannot propagate outside the unit. Testing is done by filling the enclosure space with a flammable gas and using a specially installed spark plug to ignite the gas. For core-mounted FADEC applications, the design must also demonstrate compliance with the standard flame test which requires maintenance of structural integrity during exposure to a 2000 °F flame for a specified time. Fuel cooling during this test is necessary for a successful outcome for an aluminum housing.

3.6   Concluding Commentary

The control concepts described in this chapter represent the fundamental requirements for the safe and efficient operation of the core of all gas turbine engines in both thrust and shaft power generation applications. This basic building block of the gas turbine propulsion system is referred to as the gas generator (or gas producer). With the appropriate control and protection of this machinery during transient operations as described above, the 'outer loop' control of thrust and power can be safe and effective.

Chapters 4 and 5 address how these 'outer loops' are designed and implemented within the constraints of the gas generator control.

References

1. Saravanamuttoo, H.I.H., Rogers, G.F.C, and Cohen, H. (1951–2001) *Gas Turbine Theory*, 5th edn, Pearson Education Ltd.
2. Langton, R. (2006) *Stability and Control of Aircraft Systems*, John Wiley & Sons, Ltd, UK.
3. Maccallam, N.L.R (1969–1970) The performance of turbojet engines during the thermal soak transient. *Proceedings of the Institution of Mechanical Engineers*, **184** (Part III-G).
4. Maccallam, N.R.L (1979) Thermal Influences on Gas Turbine Transients. Effects of Changes in Compressor Characteristics, ASME 79-GT-143.
5. Abernethy, R.B. and Thompson, J.W. (1973) *Handbook, Uncertainty in Gas Turbine Measurements*, Arnold Engineering Development Center, Arnold Air Force Center, Tennessee, TR-73-5.
6. Saravanamuttoo, H.I.H. (ed) (1990) Recommended Practices for Measurement of Gas Path Pressures and Temperatures for performance Assessment of Aircraft Turbines and Components, Propulsion and Energetics Panel of NATO, AGARD AR 245.
7. Dudgeon, E.H. (ed) (1994) Guide to the Measurement of Transient Performance of Aircraft Turbine Engines and Components, Propulsion and Energetics Panel of NATO, AGARD AR 320.

4

Thrust Engine Control and Augmentation Systems

This chapter addresses the thrust management and control aspects of gas turbine propulsion systems for engines that generate thrust directly from the gas energy produced by the gas generator section of the engine. Applications include both commercial and military propulsion systems.

Also included within this chapter is a description of the two primary thrust augmentation techniques that are used in both commercial and military propulsion system applications from a control system perspective, specifically:

- thrust augmentation from afterburning; and
- thrust augmentation from water injection.

It should be recognized that the operational thrust efficiency and capabilities of any gas turbine propulsion system can be strongly influenced by the design and implementation of both the engine inlet and exhaust system. This is particularly critical when considering supersonic applications where the engine air inlet system becomes a major performance enabler and where the inlet can contribute more thrust to the aircraft than the gas turbine engine itself under certain operational flight conditions.

These issues are addressed in Chapter 6 which covers the engine inlet, thrust reversing, and vectoring aspects of gas turbine propulsion exhausts systems.

4.1 Thrust Engine Concepts

The turbojet engine represents the simplest form of thrust generating engine. As described in Chapter 2, the high-energy gases from the gas generator are exhausted through a jet pipe and exhaust nozzle to generate thrust. At any given operating condition the majority of the thrust developed is proportional to the net change in momentum between the air entering the engine and the exhaust gases exiting the jet pipe nozzle. Thus the higher the

Gas Turbine Propulsion Systems, First Edition. Bernie MacIsaac and Roy Langton.
© 2011 John Wiley & Sons, Ltd. Published 2011 by John Wiley & Sons, Ltd.

exhaust gas velocity, the higher the thrust generated. The downside of this situation is that the higher the exhaust velocity, the higher the noise generated.

Today, the most common approach to thrust engine design for commercial applications is the turbofan. Here the majority of the gas energy from the gas generator is dissipated in low pressure (LP) turbine stages which drive a ducted fan. The fan moves a large mass of air at a relatively low velocity rather than a small mass of air and gas at high velocity as in the turbojet. The result is a much quieter engine since the fan discharge surrounds the high-velocity core exhaust.

Figure 4.1 shows examples of fan, turbine, and compressor arrangements for two common design approaches.

The two-spool example is typical of many of today's commercial engine such as the Pratt & Whitney JT-9D and PW4000 Series of engines. Here the fan and LP compressor share the same LP spool. The three-spool arrangement was first introduced by Rolls-Royce in the late 1960s with their RB-211 engine. This same tradition continues today with the company's Trent series of engines. In this design, the fan has its own power turbine and can therefore run closer to its optimum speed. In the two-spool approach, however, the fan speed is somewhat compromised by the requirements of the LP compressor. While the

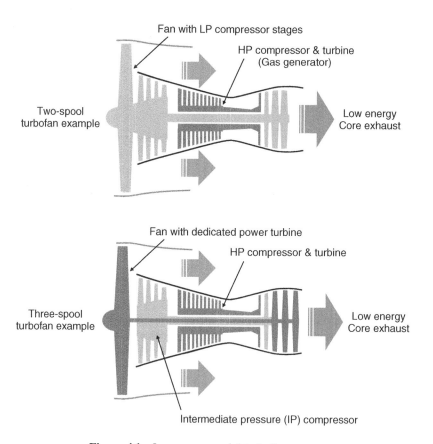

Figure 4.1   Large commercial turbofan concepts.

three-spool version is perhaps more aerodynamically optimum, the additional intermediate pressure (IP) spool adds complexity to this design solution.

Many of the smaller commercial gas turbines used in regional jet applications dispense with the LP compressor and employ a single-spool gas generator together with a power turbine to drive the ducted fan.

Military applications have important considerations that make the high-bypass turbofan unacceptable. Specifically, the large frontal area presented by the typical commercial turbofan is not compatible with the aerodynamic and stealth requirements of high-speed fighter aircraft. As a result, the military engine community has developed a series of low-profile engines that better suit the military fighter mission. Two of the most successful engine solutions that have evolved in recent years are the Pratt & Whitney F-100 and its derivatives and the General Electric F-404 family of engines.

The F-100 was developed in the late 1960s as a lightweight afterburning low-bypass turbojet for the F-15 and F-16 fighter aircraft. The General Electric F-110 was subsequently developed as an alternative power plant in order to provide effective competition in the large procurements associated with both F-15 and F-16 fleets that evolved during the 1970s and 1980s (in response to USA and foreign military aircraft market requirements).

The F-404 became the mainstay of naval fighter aircraft with the F-18 Hornet and Super Hornet, the latter of which is still in production today.

All of these engines and their derivatives are sometimes classified as 'leaky turbojets' reflecting the relatively low-bypass ratios employed (typically 0.7–0.9) which is necessary to maintain the small frontal area required.

Figure 4.2 shows the compressor/turbine/spool arrangement for the F-100 engine, which is comparable to many modern military power plants. The afterburner section and exhaust nozzle are referenced here for completeness.

As indicated in the depiction of the F-100 engine, most of today's advanced military engines employ a convergent-divergent (CD) nozzle design to provide an additional increase in thrust. With this arrangement, the exhaust gas is accelerated further through the divergent section and the pressure acting on the wall of the nozzle provides additional thrust as indicated in Figure 4.3. The CD technology used in military gas turbine

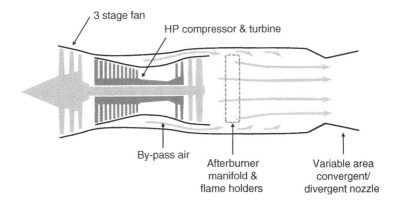

Figure 4.2   Pratt & Whitney F-100 engine spool arrangement.

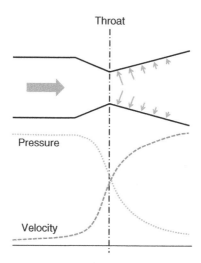

Figure 4.3   The convergent-divergent nozzle concept [1].

applications is a spin-off from rocket engine propulsion, where this type of propelling nozzle was first adopted.

A detailed discussion of thrust augmentation techniques is presented in Section 4.3 below.

4.2   Thrust Management and Control

As discussed in Chapter 2 direct measurement of engine thrust is not practical outside the engine test cell. In practice, inferred thrust measurement techniques are therefore employed to facilitate the control and management of engine thrust by the flight crew.

As previously discussed, the following three parameters are used today as a means of measuring and controlling engine thrust relative to the prevailing operating conditions:

1. engine pressure ratio (EPR);
2. integrated engine pressure ratio (IEPR); and
3. corrected fan speed: $N_1/\sqrt{T_2}$.

The first technique (which is used on Pratt & Whitney engines) uses the gas generator pressure ratio, typically P_5/P_2, where P_5 is the jet pipe pressure (P_6 for a twin-spool engine) and P_2 is the compressor inlet pressure. This parameter is a measure of the airflow through the core section of the engine.

The second technique, first introduced by Rolls-Royce on their high-bypass ratio turbofans, combines the core EPR with the fan pressure ratio algebraically. This approach is arguably more representative of the engine thrust since the majority of the engine thrust is delivered by the fan.

The General Electric Company uses the third thrust measurement technique, which is based on the fact that the airflow through the fan is a measure of total engine airflow which is proportional to corrected fan speed $N_1/\sqrt{T_2}$.

It is interesting to note that on some Pratt & Whitney engines, corrected fan speed is used as a back-up for thrust management should EPR become unavailable.

In many applications, engine sea-level static thrust is typically flat rated to some pre-determined air inlet temperature. Since airflow through the engine and hence thrust is proportional to inlet total pressure and inversely proportional to the square root of absolute inlet total temperature, the thrust generated by the gas turbine engine varies considerably over the operational flight envelope.

Figure 4.4 shows how flat-rating engine thrust at sea-level static conditions requires lower maximum throttle settings for progressively lower inlet temperatures. Also shown in the figure is the thrust lapse rate effect which combines the thrust loss with altitude and the simultaneous thrust increase as temperature decreases with altitude. After the tropopause at 36 000 ft the lapse rate increases since the temperature remains constant above this altitude (assuming a standard atmosphere).

The effect of aircraft forward speed further modifies the lapse-rate effect as both the inlet total pressure and total temperature increase with Mach number according to the following equations for adiabatic flow:

$$T_T = T_S \left(1 + \left(\frac{\gamma - 1}{2} \right) M^2 \right) \tag{4.1}$$

$$P_T = P_S \left(1 + \left(\frac{\gamma - 1}{2} \right) M^2 \right)^{\gamma/(\gamma-1)} \tag{4.2}$$

where subscripts T and S in the above equations denote total and static values, respectively. The term γ is the ratio of specific heats which is nominally 1.4 for air.

In the real world a recovery factor may be used and allowance made for additional inlet duct pressure losses.

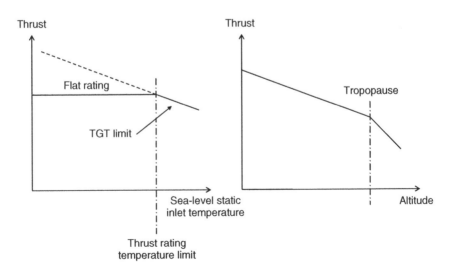

Figure 4.4 Engine thrust flat rating and altitude lapse rate effects.

There are a number of thrust settings used by commercial aircraft in service today. These are listed below, together with relevant comments as to their purpose.

Takeoff thrust. This setting defines the maximum rated thrust which is typically defined by a limiting turbine gas temperature (TGT) condition. The time allowed for operation at this condition is usually limited to about 5 minutes.

Maximum continuous thrust. This rating is determined by the engine failure on takeoff condition (after V1 has been exceeded) where the remaining engine (or engines) must be able to support a continued takeoff and climb-out. As implied by the title, the engine may be operated continuously at this thrust setting.

Maximum climb thrust. This optional setting is determined by the aircraft climb performance requirements and may be a trade-off between engine maintenance goals (TGT preferred operating limits) and aircraft operational performance.

Flight idle thrust. This setting is established by the need to provide a timely acceleration to maximum thrust initiated by a go-around maneuver. It is typically above 70% of maximum gas generator speed so that there is a greatly reduced probability of an engine surge or stall in response to a sudden large increase in throttle setting.

Ground idle. The ground idle setting is the minimum gas generator speed used on the ground during starting and taxi and represents the minimum operational thrust for the engine.

Ideally, engine thrust should be a linear function of throttle position, often referred to as power lever angle (PLA), to provide the required thrust rating at the same throttle position independent of the operational flight condition.

Prior to the introduction of electronic engine controls, the flight crew would be required to maintain the necessary thrust rating by adjusting the throttle setting as the engine inlet conditions changed following takeoff and throughout the climb to altitude. This process adds a significant workload to the flight crew and, as a result, occasional exceedances in thrust rating or TGT can occur to the detriment of the engine.

In modern fuel control systems, thrust settings for each of the above ratings are computed automatically using look-up tables based on the current flight and engine inlet conditions. Throttle settings corresponding to any required rating are therefore managed by the fuel control system which trims the gas generator speed governor set point until the correct rating is obtained, thus relieving the flight crew of additional workload and allowing them to concentrate on flying the aircraft.

Figure 4.5 shows a block diagram of this concept where PLA represents a percentage of maximum (takeoff) thrust required. The prevailing flight conditions defined by the measured engine inlet parameters are used to generate a specific thrust rating that corresponds to the PLA set by the flight crew. The concept shown is representative of both the EPR and fan speed thrust management control modes.

Note that the measured engine inlet pressure is typically static pressure rather than total pressure; Mach number information is therefore required to calculate the equivalent total pressure for the prevailing flight condition.

The above discussion on thrust management and control relates almost exclusively to commercial engine applications where engine reliability, maintainability, and cost of ownership are major drivers in operational procedures.

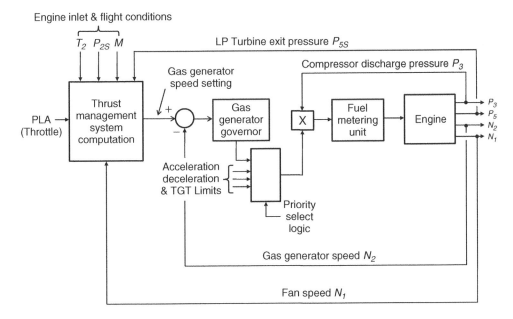

Figure 4.5   Thrust management system block diagram.

Military aircraft engines, and particularly those in fighter aircraft, see a completely different operational environment where available performance is the main driver. These engines are subject to a high number of throttle cycles in service and hence much higher mechanical and thermal stresses than their commercial counterparts, with a subsequent effect on engine life. High-cycle and low-cycle counts occurring in service are measured and used to establish engine maintenance requirements.

Pilots operating military aircraft expect maximum available thrust to be delivered at maximum PLA, which is termed 'military thrust setting' for non-augmented engines. The available thrust will, of course, vary in accordance with the prevailing flight condition. For additional thrust, augmentation techniques must be employed and these are described in the following section.

4.3   Thrust Augmentation

The benefits of thrust augmentation are obvious in military applications since higher thrust translates into higher performance. Thrust augmentation for commercial applications is desirable to maintain a specific thrust rating when operating under high air inlet temperature conditions, which limit the available thrust to that defined by TGT limits (see Figure 4.4).

There are two primary techniques that are employed to augment the available thrust: water injection and afterburning (also referred to as 'reheat'). Each of these techniques is described in the following subsections.

4.3.1   Water Injection

Available thrust is a key parameter in takeoff performance. Since air mass flow through the engine is perhaps the most important contributor to engine thrust, high air temperature which results in a reduction of air density will cause a reduction in available thrust. This loss of thrust can be recovered by as much as 30% by injecting a mixture of water and alcohol (usually methanol) into the inlet of the engine. The added alcohol provides antifreeze properties as well as providing an additional source of fuel.

It should be mentioned at this point that water injection also applies to shaft power engine applications since the result of water injection is to increase the available gas horsepower output of the gas generator.

Figure 4.6 shows qualitatively the effect of water injection on available takeoff thrust for a jet engine and available takeoff shaft horsepower for a turboprop engine.

For axial flow compressor engines, water injection directly into the combustion chamber is more suitable since it provides a more even distribution and allows greater quantities of coolant to be used. With this technique, the mass flow through the turbine section is increased relative to the mass flow through the compressor. The pressure drop across the turbine is therefore reduced, resulting in an increase in jet pipe pressure and hence thrust.

Figure 4.7 shows a schematic diagram of a high pressure (HP) type of water injection system. HP water/alcohol mixture is generated via an air turbine-driven pump and fed to a water control unit comprising a shut-off valve which remains closed until some predetermined value of compressor discharge pressure has been reached. When the valve is opened, a micro-switch senses the change in position and sends this information to the fuel system, which resets the fuel control gas generator governor to provide additional fuel flow and hence higher thrust. In this type of system, the water may be injected using dedicated passages within each of the fuel combustion nozzles. A check valve located within the water control unit prevents back flow of fuel or hot gases from the combustor.

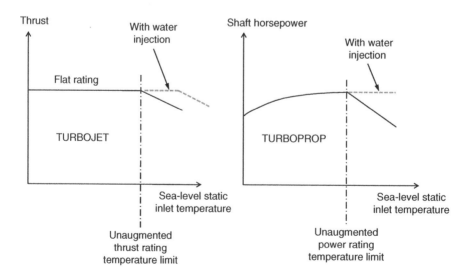

Figure 4.6   Effects of water injection on rating limits.

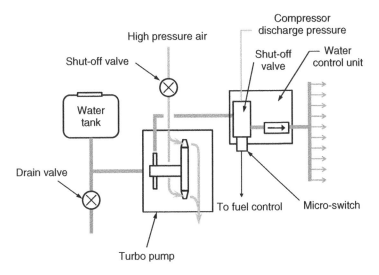

Figure 4.7 Water injection system schematic.

Water injection as a method of thrust augmentation was in common use in many of the early jet engine applications. Some of the early B-52 engines, for example, had a water injection system that employed both compressor inlet and combustor water injection.

Water injection systems are relatively heavy and therefore impose a significant operational penalty. Very few of the engines in service today use this type of thrust or power augmentation.

4.3.2 Afterburning

Since the exhaust gases from gas turbine engines contain a substantial amount of uncombined oxygen, it is clearly possible to burn additional fuel in the jet pipe in order to increase the temperature and velocity of the exiting gases hence increasing engine thrust.

Since the rate-of-change of momentum term in the overall thrust equation is related to the exhaust gas velocity, we can increase this term by increasing the temperature of the gas exhaust; this increases the associated speed of sound which occurs at the exit plane of the nozzle.

If the exhaust gas temperature is doubled, say from 1800 to 3600 R (degrees Rankine) which is close to the actual numbers that can occur in practice, then the thrust increase is in proportion to the increase in the square root of the absolute temperature ratio. By this assessment, the increase in thrust for the above example is:

$$\Delta_{\text{thrust}} = \sqrt{\frac{3600}{1800}} = 1.414. \tag{4.3}$$

This represents a thrust increase from afterburning of about 41% over the maximum 'dry thrust' operating condition. On this basis, it is typically recognized that afterburning can increase thrust by as much as 50% over the baseline dry thrust of a typical turbojet engine.

In military applications, the benefits of afterburning provide a substantially higher thrust capability without the need to increase engine size (which has an associated increase in weight and engine frontal area). Bypass engines with multi-zone afterburning can achieve an even higher thrust increase; however, the ultimate limit to this process is driven by the inability of jet pipe materials to withstand the extremely high temperatures that are generated.

When considered along with aerial refueling, the afterburner is an important force multiplier. It allows fighter aircraft to take off with a maximum weapons load from relatively short airfields, followed by a climb to altitude for rendezvous with a tanker. The fighter can continue its mission from this point with full fuel tanks.

The design features of the afterburner are described in the following sections.

4.3.2.1 Fuel Spray Bars and Flame Holders

Fuel is delivered through spray bars consisting of one or more concentric rings located downstream of the LP turbine. Further downstream, flame holders are required to form localized turbulent eddies allowing the establishment of a stable flame. This additional hardware within the jet pipe does incur some small performance penalty when operating in the dry (non-augmented) mode of operation.

4.3.2.2 Modified Jet Pipe Design

To accommodate the increased gas flow and higher temperatures that occur during after-burning, the jet pipe diameter will be somewhat larger. A cooling liner is also required to protect the jet pipe from the high exhaust gas temperatures that can be as high as 3600 R (2000 K). The burners are arranged so that the highest gas temperatures occur along the axis of the jet pipe, thus allowing a portion of the cooler LP turbine gases to flow along the wall of the jet pipe.

An additional function of the jet pipe liner is to prevent 'screech', a phenomenon that involves high frequency pressure fluctuations resulting from the combustion process that can produce excessive noise and vibration to the point where physical damage to the afterburner components can occur. The screech liner contains thousands of small holes which absorb and attenuate the thermal energy thus minimizing these fluctuations.

4.3.2.3 Variable Area Nozzle

All afterburning engines require variable area exhaust nozzles to accommodate the increased gas flow without increasing jet pipe pressure, since this can otherwise result in stalling of the HP compressor. To explain this point, consider the effect of increasing jet pipe pressure. The pressure drop across the turbine section is now reduced together with the torque delivered to compressor. The gas generator fuel control now responds by increasing fuel flow in order to maintain HP spool speed and hence the air mass flow through the engine. Thus the HP compressor operating point has moved along a constant speed line toward the stall boundary. (This point is covered in Chapter 2.)

Many early augmented engines used two-position nozzles to accommodate this requirement; however, most of today's engines employ variable area nozzle designs to allow a range of thrust-modulation during afterburning.

4.3.2.4   Afterburner Ignition System

The temperature of exhaust gases in the jet pipe is not always sufficiently high to ensure ignition of afterburner fuel, particularly at high altitudes; some form of ignition system is therefore required. Ignition of the afterburner fuel can be achieved using several different methods. The 'hot streak ignition' method uses an additional slug of fuel injected through one of the combustion nozzles into the main combustor. This generates a streak of burning fuel that passes through the turbine section of the engine to ignite the afterburner fuel in the jet pipe.

An alternative approach is a 'torch' or 'catalytic' igniter located close to the spray bars and flame holders using its own pilot fuel flow source and electrical spark initiation. Once the torch igniter device is started, it operates continuously during afterburner operation. Continuous electric spark ignition is also used in some applications, which functions in a similar manner to the torch device. Figure 4.8 shows a simplified afterburner arrangement illustrating the points discussed above.

4.3.2.5   Afterburner Control

The typical afterburner fuel control system comprises a dedicated fuel pump, a fuel control and metering unit and a nozzle area actuation arrangement. The fuel pump may employ either a positive displacement or a centrifugal pumping element since the maximum delivery pressure required for injection into the jet pipe is much lower than for the main fuel control.

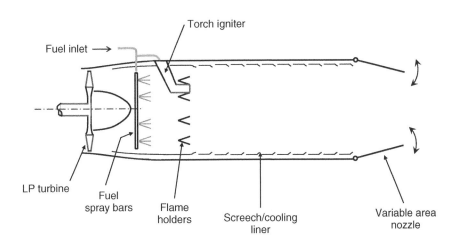

Figure 4.8   Typical afterburner arrangement.

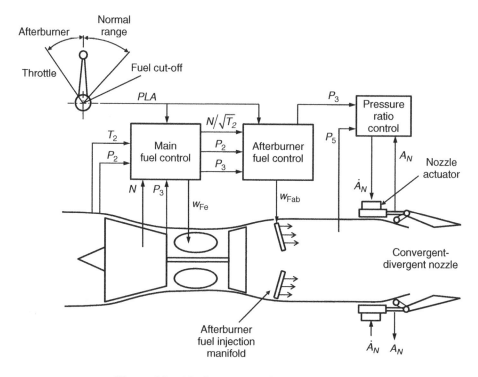

Figure 4.9   Afterburner control system overview.

The afterburner is initiated by moving the throttle lever (PLA) beyond the maximum dry thrust position. This movement signals the afterburner fuel control to fill the fuel manifolds within the jet pipe as quickly as possible, before setting the correct light-off fuel flow corresponding to the prevailing flight condition. An interlock inhibits afterburner selection if the gas generator speed (and hence airflow through the engine) is lower than some predetermined minimum. If the pilot slams the throttle from flight idle to maximum afterburner, the engine will therefore accelerate initially under the control of the gas generator acceleration limiter until the engine spool speed is sufficiently high to satisfy the interlock requirement and allow selection of the afterburner.

Once the afterburner has been ignited, fuel flow is increased until it matches the prevailing PLA setting. In the example depicted by Figure 4.9, afterburner fuel flow is determined as a function of throttle lever angle and the airflow though the engine. The latter is computed as a function of compressor pressure ratio P_3/P_2 and gas generator speed $N_1/\sqrt{T_2}$.

The exhaust nozzle is simultaneously varied to maintain a constant pressure ratio between P_3 and the jet pipe pressure P_5, thus ensuring that the compressor operating point is maintained essentially constant for all afterburner throttle settings. To accomplish this, the measured pressure ratio is compared with a predetermined nominal value and the error generated is used to drive the servo valve of the nozzle actuator; this then moves at a rate proportional to the pressure ratio error until the required pressure ratio is obtained.

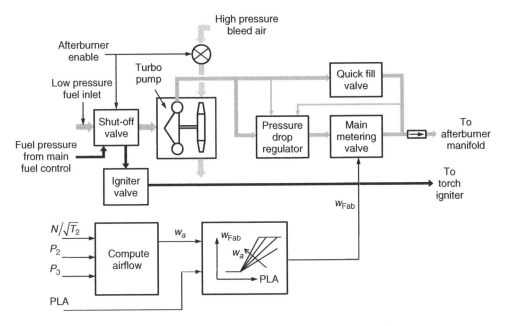

Figure 4.10 Simplified afterburner fuel control schematic.

Figure 4.10 provides a simplified schematic of the afterburner fuel control. As shown in the figure, an air-turbine driven centrifugal fuel pump provides pressurized fuel to the fuel metering section of the control. A shut-off valve powered by HP fuel from the main fuel control is maintained in the closed position until both throttle position and minimum engine speed conditions have been satisfied and the afterburner is enabled.

Upon selection of the afterburner, the first task of the fuel control is to fill the fuel manifold as quickly as possible. The quick-fill valve allows maximum fuel flow from the pump to be delivered to the afterburner manifold typically for 1–2 s before the quick-fill valve closes and the main metering valve takes over. Pressure drop across the main metering valve is maintained via a pressure regulating valve which throttles the output of the centrifugal pump accordingly. Metering valve position is therefore proportional to afterburner fuel flow.

The fuel flow command to the metering valve is a function of airflow and throttle position (PLA). This arrangement provides good handling characteristics in afterburner throughout the operating envelope of the engine.

Retarding the throttle into the normal (dry) operating range deselects the afterburner. The fuel and air shut-off valves are then closed and the fuel manifold drains overboard via a simple spring-loaded valve which opens following depletion of manifold pressure. This prevents coking of any fuel that would otherwise be retained within the manifold.

Since the afterburner fuel control is essentially a scheduling type of function, there are no closed-loop stability control issues to address. In addition, the nozzle area actuator bandwidth is much faster than the fuel control system. The closed-loop dynamics of this system are not typically a performance driver.

The example described above is perhaps an oversimplification compared to real-life applications; however, the principles presented remain applicable and appropriate from a systems perspective.

An important point to note regarding the above example is that afterburner control for the pure turbojet application is greatly simplified as a result of the choked flow through the turbine section which effectively prevents pneumatic transients within the jet pipe from influencing the aerodynamics upstream of the turbine section. This is no longer true for the more modern military engines with bypass flow around the HP section of the engine. Consideration must be made within the fuel control system to ensure that pressure disturbances within the jet pipe that propagate upstream along the bypass duct do not seriously impact the performance of the engine's HP compressor. This is typically accomplished by modulating guide vanes within the fan/LP compressor section of the engine to control airflow through the bypass duct.

4.3.2.6 Nozzle Actuation Technology

The variable area nozzle typically contains a ring of interlocking flaps enclosed by an outer shroud. Several linear actuators located around the periphery of the nozzle are used to vary the area of the nozzle under the direction of the afterburner control system. These actuators are protected from the extreme heat from the afterburner duct by an insulating blanket. Both hydraulic and pneumatic servo-actuators have been used to position the nozzle flaps. The hydraulic source may be either engine lubricating oil via a dedicated

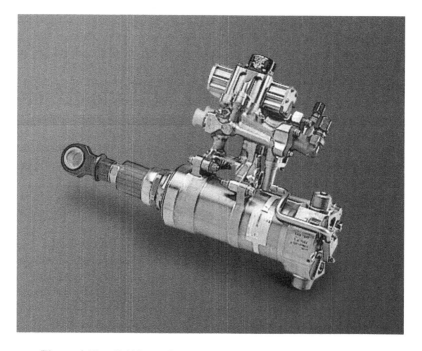

Figure 4.11   F-119 nozzle actuator (courtesy of Parker Aerospace).

HP pump or by the use of HP fuel from the fuel control system (i.e., fuel-draulics). With this approach, a continuous flow of fuel through the actuators is necessary to ensure that excessive fuel temperatures and subsequent coking does not occur. Figure 4.11 is a photograph of a fuel pressure operated nozzle servo-actuator used on the PW-119 engine which powers the F-22 Raptor aircraft.

An alternative to the hydraulically actuated nozzle actuators is used on the Rolls-Royce RB-199 engine, which powers the Panavia Tornado fighter aircraft. Here pneumatically powered nozzle actuators use HP bleed air to drive an air motor. A flexible drive from the air motor drives a number of screw-jack actuators located around the jet (similar to many thrust reverser actuating systems). This approach provides inherent synchronization of the actuator set.

Reference

1. Rolls-Royce (1969–1973) *The Jet Engine*, Rolls-Royce Ltd, Part 9.

5

Shaft Power Propulsion Control Systems

This chapter addresses the control of turboprop and turboshaft applications where gas generator power is used to generate shaft power for turboprop and helicopter applications. Auxiliary power units (APUs) also fall into this category where shaft power is used to drive electrical generators and air compressors.

One approach is to connect the gas generator shaft directly to the load via a reduction gearbox as in the Rolls-Royce Dart engine described briefly in Chapter 1. This type of gas turbine engine is sometimes referred to as a single-shaft engine. Many APUs use this type of design.

The alternative to direct mechanical power extraction is to absorb the gas generator gas horsepower in a separate turbine referred to as a 'free turbine' or 'power turbine'. This alternative approach is perhaps the most common approach used today in turboprop applications and is the only arrangement used in helicopter gas turbine propulsion systems.

In all turboprop applications, the variable pitch propeller uses a 'constant speed unit' (CSU) to control propeller speed (usually in the range 1200–2000 rpm).

Figure 5.1 shows the general concept of a gas turbine turboprop propulsion system. In both the single-shaft and free turbine turboprop engines, there are two control levers which are connected to the fuel control and the CSU: the power lever and the condition lever.

During normal flight operations, the power lever determines the power delivered to the propeller by modulating fuel flow to the engine while the condition lever determines the required propeller speed. The CSU varies the propeller pitch to maintain the requested propeller speed. By retarding the power lever below the idle setting, the pilot can control the pitch of the propeller directly in what is referred to as the 'beta control mode'. This mode is used on the ground to generate reverse thrust following touch-down and during ground operation. Since the CSU is no longer controlling propeller speed in the beta control mode, the condition lever position is now used to set an 'underspeed governor' within the fuel control to ensure that sufficient power is being delivered to the propeller as increasing negative pitch angles are applied.

Gas Turbine Propulsion Systems, First Edition. Bernie MacIsaac and Roy Langton.
© 2011 John Wiley & Sons, Ltd. Published 2011 by John Wiley & Sons, Ltd.

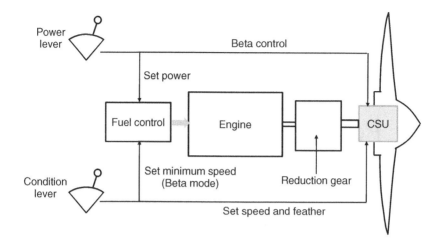

Figure 5.1   Conceptual block diagram of a turboprop propulsion system.

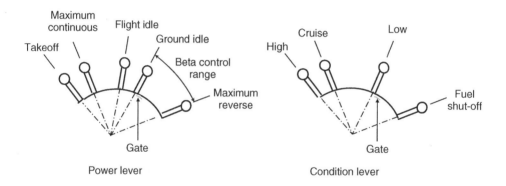

Figure 5.2   Power lever and condition lever settings.

   To illustrate the control concepts described above in more detail, Figure 5.2 depicts the various power lever and condition lever settings. Figure 5.3 depicts a functional block diagram of a typical CSU showing how the power lever and condition lever inputs contribute to the CSU modes of operation.

   Referring to Figure 5.2, the power lever is set at ground idle for starting and between flight idle and takeoff during normal flight operation. A gate at the ground idle position prevents inadvertent movement of the power lever into the beta control range.

   There are typically three condition lever settings: HIGH, CRUISE, and LOW. The LOW speed setting is used during taxiing with the power level set at or near flight idle (typically 60–75% rpm). Propeller speed is set according to the flight condition. CRUISE is typically 95–97% rpm. The HIGH speed setting (100% rpm) is used during takeoff and landing.

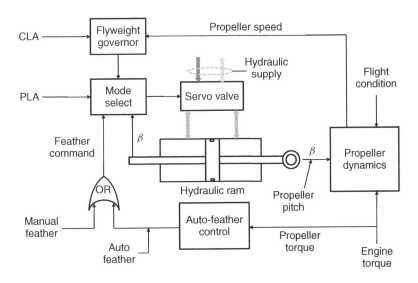

Figure 5.3 CSU functional block diagram.

Referring to Figure 5.3, the primary function of the CSU is to control propeller speed to a constant value determined by the condition lever input, condition lever angle (CLA). To accomplish this, the CSU positions a hydraulic actuator whose output is proportional to propeller pitch angle β. As pitch is increased to a more coarse position, the propeller will tend to slow down (for a given engine-delivered torque and flight condition). Any speed reduction from the required setting (CLA) will be sensed by the flyweight governor resulting in a movement of the servo valve, which in turn moves the hydraulic ram to reduce propeller pitch in order to restore the propeller speed to its nominal setting. If the power lever angle (PLA) is retarded below the ground idle setting, control of the pitch actuation system reverts to direct control of propeller pitch. In this beta control mode, further retarding of the power lever provides increasing negative pitch angles and hence higher negative thrust from the propeller.

A third mode of control provided by the CSU is the feather mode. Feathering the propeller moves the propeller pitch angle to a position of minimum drag which is particularly important if an engine failure occurs during takeoff. For this reason, many of today's turboprop engines incorporate a negative propeller torque sensor which automatically commands the propeller to feather should a negative propeller torque occur following a sudden loss of power. The auto-feather function is enabled by the pilot who can also select propeller feather manually.

The hydraulic supply to the CSU is provided by a dedicated gear pump (not shown) using oil from the engine lubrication system.

A dynamic analysis of the CSU can be accomplished using the same small perturbation linearization techniques covered in the gas generator governor response and stability analyses covered previously in Chapter 3. Root locus plots can also be developed to visualize the movement of the control system closed-loop roots as the loop gain is changed. The schematic diagram of Figure 5.4 shows the elements involved in such an analysis.

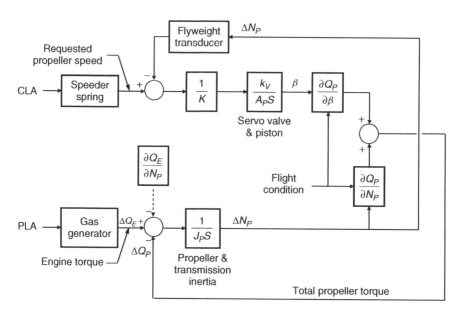

Figure 5.4 Linearized CSU block diagram: speed governing mode.

In the speed governing mode, a flyweight transducer generates a force proportional to the square of propeller speed which is compared to a 'speeder spring' force determined by the position of the CLA. In conjunction with the mechanism spring stiffness, the force difference determines the displacement of the hydraulic servo spool valve and hence the flow of oil into (and out of) the beta actuator servo piston. Propeller pitch will continue to change until the force generated by the flyweight transducer matches the force of the speeder spring set by CLA.

Since the servo valve and piston have an integration function, steady state requires that there must be zero error between the CLA speed request and the actual propeller speed. There is therefore no speed droop and the speed governing function is isochronous.

In the lower section of Figure 5.4 the gas torque generated by the gas producer section of the engine is compared to the net propeller torque; the difference is applied to accelerate (or decelerate) the effective inertia of the propeller and its associated transmission. The integrated output of this process defines propeller speed.

The challenge associated with the linear analysis of this system is the determination of the torque partial derivatives which are dependent upon propeller speed, pitch angle, and the prevailing flight condition. In order to establish the optimum gain values, a full range of operating conditions must be analyzed. At some point, a full-range model of the propulsion system becomes more cost effective as well as being valid for larger perturbations from the operating point under study.

It is interesting to note that there is very little dynamic coupling between the gas generator and the propeller and transmission, since the propeller (and hence power turbine speed) remains fairly constant at any given operating condition. Since the coupling term which causes a change in gas torque with power turbine speed is both small and stabilizing, it is therefore convenient to neglect its effect for the purposes of initial stability analyses.

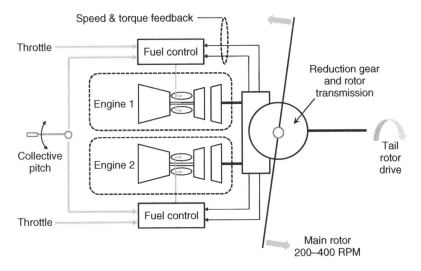

Figure 5.5   Conceptual block diagram of a twin-engine turboshaft propulsion system.

The turboshaft (helicopter) propulsion system application is fundamentally different in function from the turbojet, turbofan, or turboprop applications where the engine provides thrust to move the aircraft through air, allowing the wing to generate lift from the resulting forward speed and attitude. In the turboshaft application, the engine drives the helicopter rotor (sometimes referred to as a 'rotary wing') to generate lift directly.

Many helicopter propulsion systems employ more than one engine as indicated by the schematic of Figure 5.5, which shows a typical twin-engine helicopter rotor transmission system.

In this example, twin gas generators provide gas horsepower to two free turbines whose outputs drive a combining gearbox and rotor transmission. In the helicopter application, the rotor system is designed to operate at a constant speed of typically 200–400 rpm.

Rotor speed control is accomplished via a speed governor that modulates the gas generator fuel control from an upper limit determined by the throttle setting down to the engine idle speed or, in some applications, a minimum fuel flow limit.

In multi-engine turboshaft applications, it is important that the gas generators share the load equally. To accommodate this requirement, some applications use a torque balancing arrangement: torque from each power turbine is fed back to the fuel control and any difference is used to modify either fuel flow or gas generator governor speed. This control function is usually a slow response (low bandwidth) trim system, in order to avoid dynamic coupling with the primary control system.

The power required by the rotor system is determined primarily by the collective pitch setting. In some applications, collective pitch may be used to provide some anticipation of rotor load changes thereby minimizing any underspeed or overspeed excursions following sudden maneuvers.

One unique feature of the helicopter transmission is the need to accommodate the auto-rotation condition. While in flight following a sudden reduction in collective pitch, the resulting descent-induced airflow through the rotor will generate an acceleration torque being applied to the rotor. A clutch in the transmission system allows the rotor to separate from the engines if transmission torque goes negative, while the speed governors maintain the power turbines at the nominal governed speed.

This state is referred to as the 'split-needle' condition based on the cockpit instrument that displays both engine power turbine speed and rotor speed. While these two speeds are normally identical (taking into account the reduction gear ratio), in this state the indicator needles separate from each other.

A subsequent increase in collective pitch in the split-needle situation will reduce rotor speed. When it drops to the prevailing engine speed, the clutch will re-engage and the transmission system will revert to its normal configuration.

The following sections describe the control aspects associated with both the turboprop and turboshaft propulsion systems.

5.1 Turboprop Applications

5.1.1 The Single-shaft Engine

As described briefly above and in Chapter 1, the single-shaft engine speed is governed by the propeller CSU which modulates propeller pitch to absorb any gas generator shaft power over and above that required to drive the compressor. As fuel flow is increased, pressures and temperatures within the gas generator will increase while speed is maintained constant. With this type of engine, management of fuel flow during throttle transients is somewhat different from the traditional acceleration and deceleration limiting techniques described in Chapter 3, where the gas generator fuel flow is limited as a function of gas generator speed and other parameters. One of the most popular single-shaft turboprop engines in service today is the Honeywell TPE331 which is shown is the functional schematic of Figure 5.6.

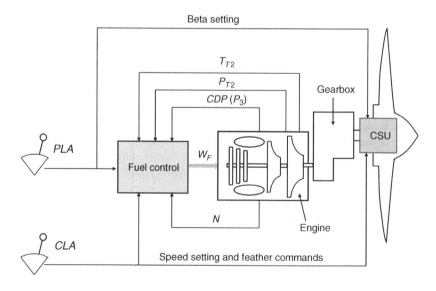

Figure 5.6   TPE331 single-shaft turboprop schematic.

As indicated in the figure, the engine comprises a two-stage centrifugal compressor driving a three-stage turbine. The engine delivers about 900 SHP at a maximum speed of approximately 42 000 rpm. The gearbox reduces engine speed to 2000 rpm at the propeller.

The fuel control modulates fuel flow to the engine as a function of PLA, compressor discharge pressure (CDP), and compressor inlet pressure P_{T2} and temperature T_{T2}.

A separate CLA provides a speed set point to the CSU at the propeller. The condition lever also provides a manual feather command to the CSU.

To illustrate the control concept of this particular single-shaft engine application in more detail, the schematic of Figure 5.7 shows a simplified version of the current Woodward Governor fuel control.

The high pressure (HP) fuel pump is of traditional design, comprising a centrifugal backing pump together with an HP gear stage. Also included in the HP pump unit, but not shown in the schematic, are a relief valve, a filter with bypass arrangement, and an anti-ice valve which ports warm fuel to the pump from the fuel/oil heat exchanger.

The control section of the fuel control comprises a 3D cam driven axially by an engine inlet total pressure sensor and rotationally by a CDP actuator. Cam followers combine with the throttle input (PLA) and inlet total temperature to determine the fuel flow allowed for the prevailing flight and engine power condition.

A 'select high' linkage arrangement allows the underspeed governor (2) to maintain a minimum engine speed during beta mode operation. During normal flight operation, this underspeed governor does not come into play.

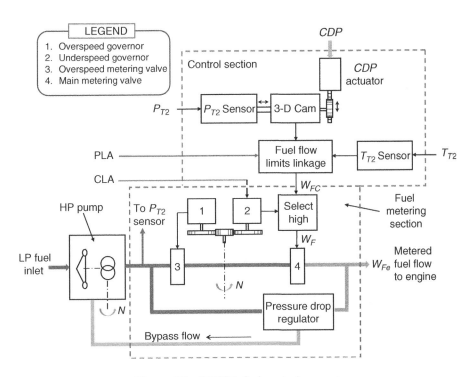

Figure 5.7   TPE331 fuel control concept.

The main metering valve (4) is positioned by the control section linkage and the pressure drop across the metering valve is maintained constant by a fuel bypass pressure regulator.

The position of the main metering valve is therefore proportional to metered fuel flow. Metered fuel passes through a pressurizing and shut-off valve (not shown) on the way to the engine combustion nozzles. A separate overspeed governor (1) controls its own dedicated metering valve (3) located in series with the main metering valve. During normal operation, the overspeed governor metering valve will be wide open as the main metering valve controls fuel flow. If an overspeed occurs, the overspeed metering valve will take over control from the main metering valve thus preventing any dangerous engine overspeed from occurring.

Proponents of the single-shaft engine prefer the fast power response to power lever changes. Since the gas generator speed is held constant by the CSU, changes in power output occur quickly as internal pressures and temperatures change. In the free turbine engine, changes in power setting must wait for the gas generator speed to reach its new steady-state value before the required power is achieved.

5.1.2 The Free Turbine Turboprop

As mentioned in several places in this book, one of the most successful gas turbine engines of the past 50 years or so is the Pratt & Whitney Canada's PT-6 free turbine engine which is used in many different turboprop and turboshaft applications. While this engine is an excellent example of free turbine engine technology, it is a mature design with a fuel control system based on the technologies of the 1960s (as is the TPE331 covered in the previous section). It is therefore considered more appropriate to include a more modern free turbine turboprop engine as an example for discussion here.

One of the most recent free turbine turboprop engines, which was developed as a family covering a wide range of power requirements, is the PW100 series (also manufactured by Pratt & Whitney Canada). The PW150A, the most recent of the series, was developed to power the Bombardier Dash8 Q-400 aircraft which entered service in 2000 and develops about 5000 SHP. The PW100 series is considered growable to about 7000 HP which would be almost an order of magnitude more powerful than the PT-6.

The PW150A engine is depicted in functional diagram and photograph form in Figures 5.8 and 5.9, respectively, the former showing the spool and airflow arrangement. This engine is more conventional in its physical arrangement (Figure 5.8) with the air inlet below the reduction gearbox at the front. With the two-stage free turbine located aft of the gas generator, the drive shaft to the reduction gear and propeller must pass through the center of the gas generator spools.

In this application, the conventional all-hydromechanical CSU is replaced by an electronic-hydromechanical arrangement comprising a propeller control unit (PCU) and associated sensors within the propeller hub and a remotely mounted propeller electronic control (PEC). A description of the propeller control system is covered in more detail below.

The PW150A has a two-spool gas generator with the low pressure (LP) compressor comprising three axial stages driven by a single-stage turbine. The HP spool has a single-stage centrifugal compressor also driven by a single-stage turbine. The reverse flow combustor helps to keep the engine length at a minimum.

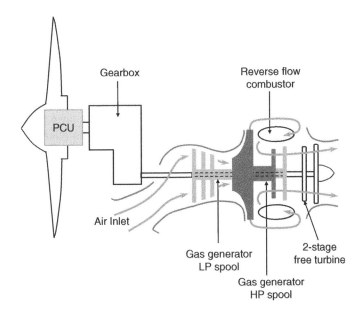

Figure 5.8   PW150 turboprop engine arrangement.

Figure 5.9   The PW150A engine (courtesy of Pratt & Whitney Canada).

There are no variable compressor stator vanes on this engine; however, there are two compressor bleed valves as follows:

- an interstage bleed valve (IBV) located at station 2.2 on the LP compressor; and
- a handling bleed valve (HBV) located at station 2.7 on the HP compressor.

The IBV remains open during starting and to prevent stalls during low-power transients. This valve is fully closed before cruise power is reached. The HBV function is to improve surge margin during rapid throttle movements.

The fuel control system comprises a modern full-authority digital electronic control (FADEC) digital electronic control interfacing with a fuel metering unit (FMU). The latter incorporates a conventional HP pump comprising a backing stage supplying a positive displacement gear stage. A simplified schematic of the FMU is shown in Figure 5.10 (the fuel/oil heat exchanger between the two pump stages is omitted for clarity). The schematic also shows a dual permanent magnet alternator (PMA) which provides dedicated power to the FADEC. The PMAs are shown since they are driven from the same auxiliary gearbox pad as the HP fuel pump.

As shown in Figure 5.10, the torque motor and associated linear variable differential transformer (LVDT) feedback allows the FADEC to control the position of the fuel metering valve. Since the pressure drop across the metering valve is controlled to a constant setting (with a temperature biasing element), metering valve position is thus proportional to metered fuel flow. Both the torque motor and LVDT have dual windings and are connected to the FADEC via individual connectors to provide the required functional integrity.

Fuel cut-off is selected through the FADEC either via a switch on the condition lever quadrant or following the detection of an overspeed condition. The fire-handle on the flight deck drives the separate solenoid valve independently of the FADEC.

In this application, the FMU also provides motive flow to the airframe to drive feed and transfer ejector pumps from the spill flow in excess of the engine fuel demand. The motive flow valve is not shown in the schematic.

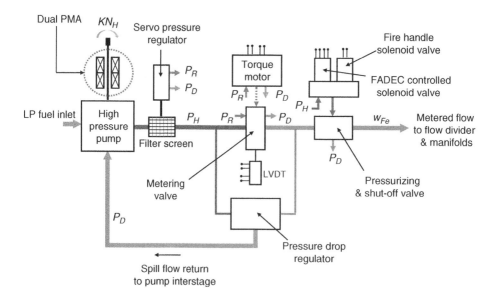

Figure 5.10   PW150A fuel metering unit schematic (courtesy of Pratt & Whitney Canada).

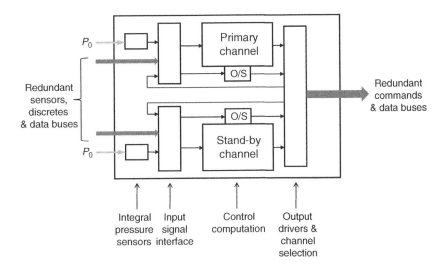

Figure 5.11   FADEC architecture.

The schematic of Figure 5.11 shows the basic architecture of the PW150A FADEC which uses dual channel master/standby architecture with internally mounted pressure sensors which measure local ambient pressure P_0. P_3 (CDP) is also provided to the FADEC from remotely mounted pressure sensors for optimum performance.

Each control channel receives both sensor sets of data and the software operated by each channel is identical. The health status of each channel is made available to the channel switchover logic which determines which channel shall remain in control. The electrical stages of all output effectors have dual windings to ensure single failure tolerance. An independent overspeed detection circuit with its own dedicated N_G speed input is also located within each channel.

Figure 5.12 shows the complete fuel and propeller control system in simplified schematic form. The system comprises a PEC which is also a dual channel microprocessor-based unit. In addition to providing the basic pitch control and speed governing functions typical of the CSU concept, the PEC provides a number of safety functions which are incorporated via an independent electronic circuit.

These safety functions are:

- auto-feather (when enabled), based on a predetermined negative torque threshold;
- auto-propeller underspeed control; and
- auto-takeoff power control up-trim command to the opposite engine.

Propeller overspeed protection is provided by a separate hydraulic overspeed governor (in the forward propeller pitch regime) and by the FADEC (for operation in reverse pitch).

An important function provided by the PEC in this application is propeller synchronization. For the Dash 8 Q-400 aircraft, the Q stands for 'quiet'; a major selling feature for this turboprop aircraft is low cabin noise. The propeller synchronizer synchronizes both speed and angular position of both propellers in order to provide an optimum cabin noise signature.

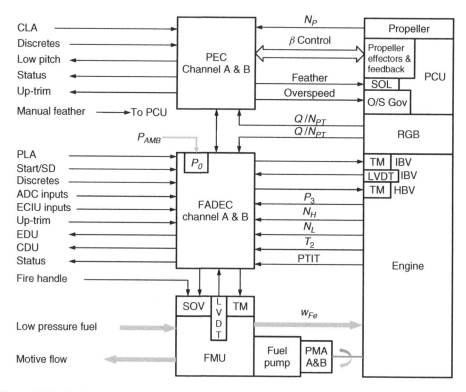

Figure 5.12   Engine fuel and propeller control system overview (courtesy of Pratt & Whitney Canada).

In order to simplify the schematic of Figure 5.12, a number of sensors and functions have been omitted for clarity. These include FADEC control of the ignition system as well as monitoring of the lubrication system chip detectors, oil pressure, and temperature. Power supply details associated with the PEC and FADEC are shown separately in Figure 5.13 and have been omitted from Figure 5.12 to reduce diagram complexity.

5.1.2.1 Power Rating System

The PW150A control system incorporates a comprehensive power rating system that greatly reduces pilot workload, allowing the crew to focus on flying the aircraft. There are two ways to set a power rating, described below.

1. By setting the condition lever (CLA) to a specific propeller speed setting. Default engine ratings are selected as follows:

    a. setting CLA to 1020 rpm automatically sets normal takeoff power (NTOP);
    b. setting CLA to 900 rpm automatically sets maximum climb (MCL) power; and
    c. setting CLA to 850 rpm automatically sets maximum cruise (MCR) power.

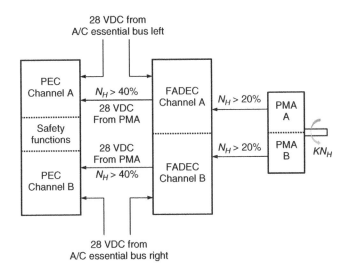

Figure 5.13 FADEC and PEC power supply architecture (courtesy of Pratt & Whitney Canada).

2. By selecting push-buttons for MCL or MCR, which will override the default CLA position rating for MCL and MCR. Climb and cruise power ratings can therefore be achieved at both the 900 and 850 rpm propeller speeds.
3. By selecting the push-button for maximum takeoff power (MTOP), which will override the default CLA position rating for NTOP. The pilot-selectable maximum takeoff power (MTOP) is available at the 1020 rpm propeller speed with PLA at the rating detent.

Figure 5.14 which shows how PLA relates to the various power ratings. An emergency power setting is available which provides up to 125% of MTOP by advancing the power lever (PLA) into the over-travel region as shown in the figure. However, this power lever setting is only intended for use in extreme circumstances.

The PW150A FADEC control system schedules fuel so that the power delivered is a linear function of PLA. When the PLA is in the 80° detant position, the crew can select different power ratings which are achieved automatically without the need for adjustment of PLA.

5.1.2.2 The PW150A FADEC Fuel Control

The FADEC incorporates a sophisticated fuel control system which provides closed-loop control of power as a function of PLA, in combination with the prevailing operating conditions which include:

- PLA and CLA inputs;
- engine rating and engine environmental control system (ECS) bleed selections;
- engine inlet conditions;
- indicated airspeed (from the aircraft air data computer or ADC); and
- up-trim from the remote PEC if applicable.

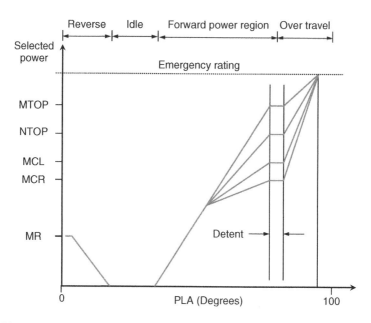

Figure 5.14 Power rating scheme (courtesy of Pratt & Whitney Canada).

Air data information (via aeronautical radio incorporated or ARINC 429) together with FADEC and engine sensors are used to determine the aircraft flight condition. FADEC and engine sensors have priority over the ADC inputs, with the exception of indicated airspeed which is only available from the ADC.

The FADEC fuel control logic involves three nested control loops as follows:

1. an outer loop providing closed-loop control of power;
2. an intermediate loop providing closed-loop control of gas generator speed; and
3. an inner loop providing closed-loop control of fuel flow.

The outer loop provides closed-loop control of engine shaft horsepower for PLA settings at or near the selected rating.

The intermediate loop controls gas generator speed N_H, with loop constraints associated with N_L limiting, w_F-dot limiting, N_H acceleration, and deceleration limiting.

The inner loop controls fuel flow and hence the position of the metering valve in the FMU. This loop incorporates a dedicated engine start schedule which includes a power turbine inlet temperature (PTIT) monitor which aborts the start process during ground starting but not during in-flight starting. Fuel flow and fuel flow rate limits are also provided within this control loop.

The above example demonstrates how complex control algorithms can be readily incorporated within the modern FADEC, which has enormous computing power. Nevertheless, complex software designed and qualified to Level 'A' critical standards is an expensive process; the task of delivering such a product to an acceptable level of maturity at entry into service is extremely challenging.

Major improvements in performance and fault-recording techniques are also present in modern electronics. This involves downloading performance and health status information, including detailed fault codes, to both flight-line data-gathering facilities as well as more sophisticated off-line analysis tools.

See Chapter 10 for a detailed description of prognostics and health monitoring (PHM), which has become a major issue in both the commercial and military communities as a means to radically change the approach to maintenance, logistics, and operational cost management.

For the discerning reader, [1] is an excellent description of the PW-100 series of engines.

5.2 Turboshaft Engine Applications

The engines that power rotary wing aircraft (helicopters) are referred to as turboshaft engines; in essentially all applications, the free power turbine engine arrangement is used as the means to absorb power from the gas generator.

The helicopter rotor is designed to operate at a constant speed over the full aerodynamic load range. Rotor power is determined by the collective pitch set by the pilot and the fuel control system modifies the fuel supplied to the gas generator in order to maintain the rotor at its nominal operating speed.

The task of controlling rotor speed using a power turbine speed-governing system is particularly challenging; the massive inertia of the main rotor which produces a relatively low resonant frequency mode can seriously impact the vehicle ride quality. This resonance can also play a significant role in the stability of the power turbine speed governor that is at the heart of the turboshaft control system.

In the process of controlling (governing) the speed of the rotor, the fuel control system modulates the gas horsepower provided to the power turbine via two common control methodologies:

- by resetting the gas generator governor; or
- by directly modulating the fuel flow to the gas generator.

Both of these techniques are common practice in today's turboshaft propulsion applications, and there are benefits and problems associated with both of these techniques which are discussed in the following text.

First let us consider the dynamics of power turbine and rotor transmission, which is the mechanical load being driven by the gas energy output from the gas generator. In its simplest form, the rotor transmission system can be represented by the schematic of Figure 5.15 where the rotor inertia (tail and main) are lumped together as a single inertia referred to the power turbine and the transmission gearbox is represented by a single stiffness term.

There are two damping terms: the rotor damping term reflects the aerodynamic drag on the rotary wing in terms of torque change per unit rotational speed; and the second term represents the change in gas torque delivered by the engine per unit speed which is a function of the gas generator operating condition.

Even in this simplified form, the dynamics are somewhat complex. This will become clear as we develop the transfer function relating applied gas torque from the gas generator

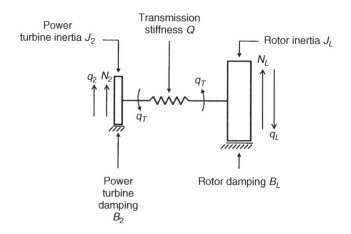

Figure 5.15 Helicopter transmission simplified.

Table 5.1 Rotor transmission parameters.

Parameter	Variable	US/Imperial units	Système Internationale (SI) units
Power turbine gas torque	q_2	ft lb	N m
Power turbine speed	N_2	rad/s	rad/s
Rotor speed	N_{L}	rad/s	rad/s
Transmission torque	q_{T}	ft lb	N m
Rotor load torque	q_{L}	ft lb	N m
Transmission stiffness	Q	ft lb/rad	N m/rad
Turbine damping	B_2	ft lb/rad/s	N m/rad/s
Rotor damping	B_{L}	ft lb/rad/s	N m/rad/s
Turbine inertia	J_2	ft lb/rad/s^2	kg m^2
Rotor inertia	J_{L}	ft lb/rad/s^2	kg m^2

to the power turbine speed in order to apply this term to our analysis of the power turbine speed governor.

Before analyzing the rotor transmission dynamics, we first review the units (in both the traditional US and SI form) for each term as listed in Table 5.1.

The torque balance and equations of motion for the above system are as follows:

$$q_{\mathrm{T}} = \text{shaft twist} \times Q = \frac{1}{s}(N_2 - N_{\mathrm{L}}), \tag{5.1}$$

$$q_2 - q_{\mathrm{T}} - B_2 N_2 = J_2 s N_2 \tag{5.2}$$

and

$$q_{\mathrm{T}} - q_{\mathrm{L}} - B_{\mathrm{L}} N_{\mathrm{L}} = J_{\mathrm{L}} s N_{\mathrm{L}}. \tag{5.3}$$

(See Table 5.1 for notation definitions.)

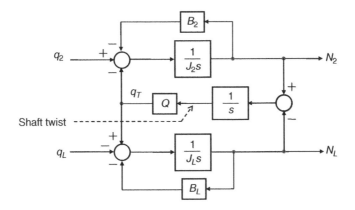

Figure 5.16 Helicopter transmission block diagram.

In order to clarify the task of developing the transfer function relating gas torque and power turbine speed, we can construct a block diagram as depicted in Figure 5.16. From this diagram we can develop the transfer function from turbine torque to turbine speed using the same small perturbation methodology that we used for the gas generator governor analysis in Section 3.2.3. In doing so, we can neglect the rotor load torque by assuming that it remains constant for the operating condition under study.

The transfer function reduces to the following:

$$\frac{\Delta N_2}{\Delta q_2} = \frac{K_L(s^2/\omega_L^2 + (2\zeta_L/\omega_L)s + 1)}{(1 + \tau_L)(s^2/\omega_n^2 + (2\zeta_n/\omega_n)s + 1)} \tag{5.4}$$

where $K_L = 1/(B_2 + B_L)$ is the load gain; $\omega_L = \sqrt{Q/J_L}$ is the rotor resonant frequency; $\zeta_L = B_L\omega_L/2Q$ is damping ratio; $\tau_2 = K_L(J_2 + J_L)$ is the load time constant; and

$$\omega_n = \sqrt{\frac{Q(J_2 + J_L)}{J_2 J_L}}$$

is the combined resonant frequency. Note that $\zeta_n = (B_L + B_2)\omega_n/2Q$.

We can now use the above load transfer function to analyze the response and stability of the power turbine governor, which must generate inputs to the gas generator that will maintain the power turbine (and hence the helicopter rotor) at a constant speed.

Let us consider now the gas generator as a 'gas torque' producer at the power turbine. The torque response follows a similar form to the CDP response in that it has 'fast-path' and 'slow-path' contributions to the total torque. The fast-path torque response is the result of the change in fuel flow and the slow-path response follows as gas generator speed changes to the new steady-state condition. Figure 5.17 shows the small signal block diagram from fuel flow to gas torque as well as how this is reduced to a gain and a lead term in series with the primary engine lag. As before, the small lags associated with the combustion process have been neglected for clarity but should be included in any rigorous analysis.

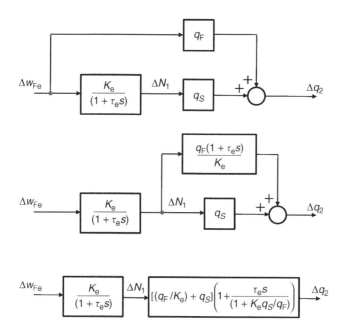

Figure 5.17 Small signal power turbine torque response.

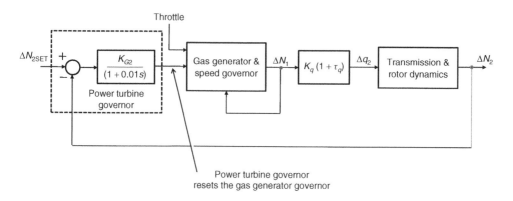

Power turbine governor
resets the gas generator governor

Figure 5.18 Power turbine governor small signal block diagram: governor reset mode.

In our analysis we can simplify the lead term in Figure 5.17 to $K_q(1 + \tau_q s)$ where $K_q = (q_F/K_e) + q_S$ and $\tau_q = \tau_e/(1 + K_e q_S/q_F)$.

Figure 5.18 depicts a complete block diagram of the N_1 governor reset concept of power turbine governing for small perturbations about a selected operating condition. Referring to the figure, the power turbine governor is represented by a simple gain with a small linear lag of 10 ms. This device resets the gas generator governor in order to maintain a constant power turbine speed. The throttle input to the gas generator governor represents the maximum power available to the propulsion system. In helicopter applications, the throttle is typically a twist grip as part of the collective pitch lever.

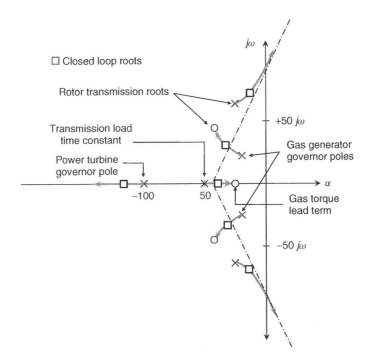

Figure 5.19 Root locus plot of a typical power turbine governor using the gas generator governor reset concept.

In our examination of the power turbine governor performance, we can use the gas generator governor closed-loop roots developed earlier in this chapter for analyzing this system. If we take a qualitative view of this closed-loop system from a stability perspective, we can generate a root locus plot (see Figure 5.19) indicating how the closed-loop roots of the power turbine governor move as the loop gain is increased.

Referring to the figure, the transmission dynamics are represented by the complex conjugate pole and zero pairs in the 45–50 rad/s region and the load time constant at −50 along the negative real axis. Damping ratios in the range 0.2–0.4 are assumed for the oscillatory roots with the rotor resonance having the higher damping. These numbers are examples and may vary considerably with the specific application; however, the observations from the root locus plot still apply.

The gas generator governor is represented by the two complex conjugate poles at 20 rad/s with good damping ($\zeta = 0.7$). The additional real root at about −150 has been neglected for simplicity since its effect will be very small. The phase advance term associated with the gas torque generated from gas generator speed changes is assumed to be at −25 on the real axis. Finally, the power turbine governor is assumed to have a 10 ms linear lag represented by the pole at −100.

As shown on the root locus plot, the loci from the complex conjugate rotor transmission poles are the limiting factor in determining the governor loop gain; the closed-loop response will inevitably contain an oscillatory component close to the rotor resonant frequency. With today's electronic control technology, a notch filter may be used to cancel the primary rotor resonance pole thereby alleviating this situation and improving achievable response and ride performance.

Many power turbine governors in service today employ droop-type governor designs which are relatively simple to implement using hydromechanical technology. In the example shown, loop gains of more than about 10.0 are limited by the transmission resonance system with complex conjugate poles which quickly move across the $j\omega$ axis as loop gain is increased. However, the achievable steady-state speed droop over the full load range is generally acceptably small (typically less than 5% of the nominal power turbine, and hence rotor, speed).

For maximum small signal bandwidth, the w_F/P_1 control mode is preferred over the w_F/P_3 alternative because the N_1 speed governor with this control mode can be faster for a given damping ratio. The former is also quite adequate for the limited altitude range typical of rotary wing aircraft.

With the gas generator governor reset approach, the achievable bandwidth of the N_2 governor is of the order 2–3 Hz. It should be noted, however, that the rotor speed response to large load excursions will be determined by the acceleration and deceleration limits associated with the gas generator. Only a large signal analysis can establish the specific power turbine (rotor) speed performance excursions for such maneuvers.

The alternative power turbine governing control mode to the gas generator governor reset concept described above employs direct trimming of the fuel flow command to the fuel metering system. Here, the power turbine governor command becomes one of the selectable gas generator control modes which take control of the engine fuel flow command if other priorities such as acceleration, deceleration, and temperature limitations allow. (Refer back to Figure 3.10 which outlines this control methodology.)

The small signal block diagram for the power turbine governor using the fuel flow trimming method is shown in Figure 5.20. As shown in this figure, the difference between the fuel flow trim and governor reset systems is that the FMU dynamics and the primary engine time constant are now in the forward path of the system. The equivalent root locus plot for the fuel flow trim system is shown in Figure 5.21. As shown in the figure, the N_1 governor poles of the previous system have been replaced by the engine primary time constant pole close to the origin (say, at -1.0) and an additional pole at -50 representing the fuel metering dynamics. The impact of this is to realize a closed-loop root at about -5.0 (representing a time constant of 0.2 s) which becomes the dominant response feature of the power turbine governor.

The good news, however, is that a good loop gain is achievable with this approach due to the attenuation provided by this closed-loop root at frequencies close to the rotor resonance.

When calculating the loop gain using the root locus technique it must be remembered that there is a change in the format of the open-loop transfer function from the traditional arrangement. For example, the root locus format of

$$\frac{K_L(1 + \tau_1 s)}{(1 + \tau_2 s)(s^2/\omega^2 + (2\zeta/\omega)s + 1)}$$

is

$$\frac{\omega^2 K_L T_1/T_2(s + 1/T_1)}{T_2(s + 1/T_2)(s^2 + (2\zeta\omega)s + 1)}.$$

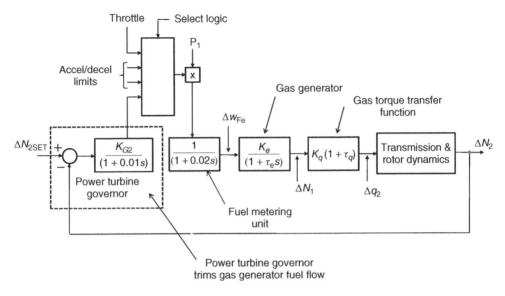

Figure 5.20   Power turbine governor small signal block diagram: fuel flow trim mode.

This means that the loop gain calculated from the root locus plot now includes the additional terms which must be accounted for in arriving at the correct value of K_L. In other words, loop gain K_L is defined as:

$$K_L = \frac{\text{(Product of vector lengths from poles to the point on the locus)}}{\text{(Product of vector lengths from zeros to the point on the locus)}} \times C_F \quad (5.5)$$

where C_F is the correction factor associated with the terms described above.

Having determined the true loop gain for the desired root positions, the value of the governor gain can be established from the contributing system component gains around the loop.

In summary, the governor reset concept may provide better small signal bandwidth than the fuel flow trimming concept even though the droop effects of both systems are quite good. Both of the above power turbine governing concepts are employed extensively in turboshaft systems in service today.

In the above analysis, we have addressed the stability of the turboshaft power turbine governor using the characteristic equation and root locus methodology to visualize how the governor loop gain affects the stability margins of the system.

It should be remembered that, in order to obtain the frequency response of the complete power turbine governing system (i.e., how power turbine speed responds to changes in the governor speed setting), we need to include the numerator terms in the forward path of the closed-loop transfer function. In other words, the closed-loop transfer function for the power turbine governor is as follows for small perturbations:

$$\frac{\Delta N_2}{\Delta N_{2SET}} = \frac{\text{Forward path numerator terms}}{\text{Closed-loop roots}}. \quad (5.6)$$

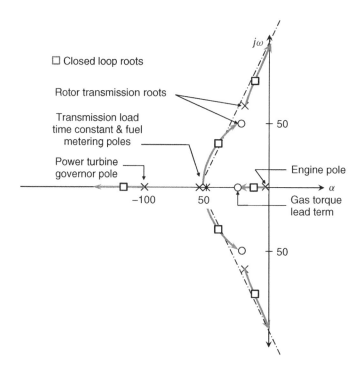

Figure 5.21   Power turbine governor root locus plot for the fuel flow trim system.

In our example, the numerator terms are the second-order system associated with the transmission dynamics and the gas torque lead term. The closed-loop roots are those developed via the above root locus analysis. The frequency domain diagram of Figure 5.22 shows the poles and zeros associated with the closed-loop response defined in Equation 5.5 above for the governor reset type of power turbine governor.

The frequency response of this system can be defined for specific frequencies (points on the $j\omega$ axis) where the amplitude ratio is the product of the vector moduli from the zeros to the point divided by the product of the poles to the same point (remember to take into account the correction factor C_F associated with the root locus format described above). Similarly, the phase angle is the sum of the angles from the zero vectors minus the sum of the pole vector angles. This is illustrated by point $P_{j\omega}$ and the associated vectors on the above figure.

In helicopter operation, however, the system dynamic response of most interest to the system designer is the ability of the power turbine governor to maintain a constant rotor speed following large changes in the rotor's aerodynamic load (specifically, sudden changes in collective pitch). All of the above analyses are based on small perturbations about a nominal operating condition; while this serves to determine the best governor gains for good response and stability, it does not address the system response to large load excursions.

In order to ascertain how the system responds to large excursions in collective pitch, we need to develop a full range system model which includes all of the relevant engine and

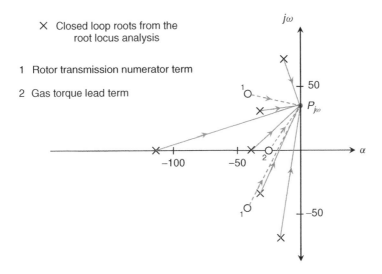

X Closed loop roots from the
    root locus analysis

1 Rotor transmission numerator term

2 Gas torque lead term

Figure 5.22 Closed-loop transfer function poles and zeros for the governor reset mode.

control system non-linearities including, for example, gas generator acceleration limits and changes in engine response characteristics with operating power level.

Engine manufacturers have proprietary models of their engines which are typically developed from detailed aerodynamic and thermodynamic data associated with compressor, turbine, and combustor designs that have been developed and tested over several generations.

A simpler approach to full-range engine modeling, sometimes used by the fuel control system community, provides an effective tool for the establishment the preliminary fuel control design concepts. This approach avoids the need for engine manufacturers to share proprietary design information with the control system supplier. Figure 5.23 is an example of such a full-range non-linear model for a single-spool gas generator driving a free power turbine together with a helicopter rotor transmission.

Figure 5.23(a) shows the fuel control system comprising the power turbine governor, gas generator control, and fuel metering system dynamics. Note that the power turbine governor shows an optional anticipatory input from the collective pitch command. This transiently resets the gas generator governor in anticipation of the up-coming aerodynamic load change, in order to minimize the resulting rotor speed droop that occurs following large load excursions. Acceleration and deceleration limits are included to provide realistic fuel flow response. The dynamics of the fuel metering system are represented by a simple time constant which is typically adequate for initial response evaluation.

Figure 5.23(b) shows a non-linear model of the gas generator which includes both gas generator turbine and power turbine net torque response as a function of the operating power level, so that a representative dynamic behavior is obtained over the full power range of the engine. For clarity, the engine model shown is valid for a given set of engine inlet conditions. However, by using corrected parameters (for speed, pressure, and temperature in the characteristic curves) the engine model can be adapted to any predetermined flight condition by inserting specific values of δ and θ into the model.

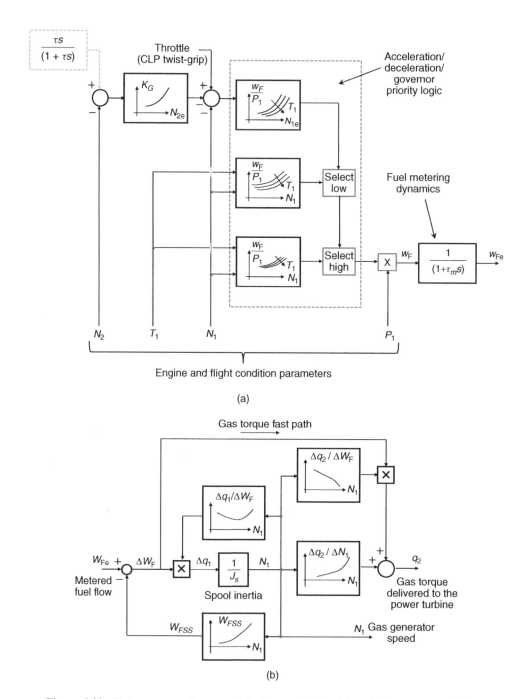

(a)

(b)

Figure 5.23   Full range non-linear model of a gas-turbine-driven helicopter transmission.

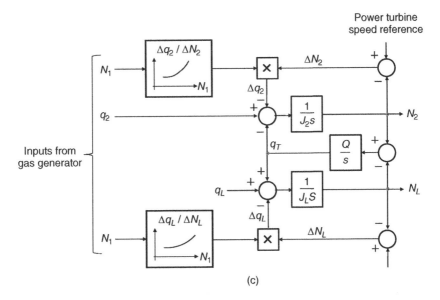

Figure 5.23 *(continued).*

Figure 5.23(c) is the large-signal equivalent of Figure 5.16 above. Here the variations in rotor and power turbine damping as a function of power level are included. The rotor torque q_L is a function of the flight condition and the prevailing collective pitch setting. The complete system response to large excursions in collective pitch can therefore be examined for various power turbine governor gains and configurations, providing a realistic indication of rotor speed response under the most demanding operational conditions.

An important performance parameter for helicopter transient behavior is the rotor speed response following a large change in collective pitch (and hence torque demand on the power turbine and transmission system). This is particularly important in military applications, where a helicopter may need to be located below the visibility horizon of the enemy but be able to 'pop up' into visual range momentarily to observe and react to the situation ahead of an enemy response. Figure 5.24 shows qualitative responses for an 80% increase in collective pitch typical of a 'jump takeoff' maneuver for three different power turbine governor gain settings.

As shown in the figure, the lowest gain exhibits significant transient underspeed with the final steady-state speed having significant speed droop from the nominal rotor speed required. The high gain response shows the high-frequency content associated with the rotor transmission resonance coming through at the start of the transient. This indicates inadequate stability margins and some significant degradation in ride quality.

As the length of this chapter indicates, the control issues associated with the turboprop and turboshaft gas generator applications are extensive compared to the thrust engines. In both cases, however, the control of thrust and shaft power is accomplished around the core gas generator control described in Chapter 3.

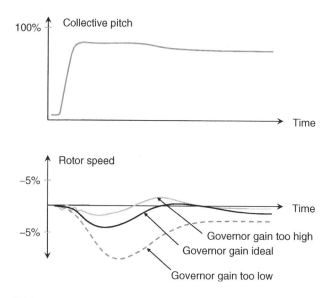

Figure 5.24   Rotor speed responses to a sudden increase in collective pitch.

Reference

1. Hoskins, E., Kenny, D.P., McCormick, I., Moustapha, S.H., Sompath, P., Smaily, A.A. (1998) The PW100 engine: 20 years of gas turbine engine technology evolution. *RTO ATV Symposium, Toulouse, France, May 1998*, Pratt & Whitney Canada Inc.

6

Engine Inlet, Exhaust, and Nacelle Systems

This chapter addresses propulsion systems issues associated with the installation of the gas turbine into the aircraft. In many military aircraft applications, stealth may be an important design driver and the engines are often buried within the fuselage or wings. In commercial aircraft applications, perhaps the most important design driver is operational cost and therefore easy access and low-maintenance expenses are important. As a result, it is typical for commercial aircraft to install their propulsion systems within nacelles mounted on struts to the wings or fuselage.

This chapter will address the following topics associated with engine installation from a systems perspective:

- engine air inlets for subsonic applications;
- supersonic air inlet systems including the Concorde air inlet control system (AICS) example;
- nacelle cowl anti-icing; and
- thrust vectoring and reversing systems including lift/propulsion engines for vertical or short takeoff and landing (V/STOL) applications.

6.1 Subsonic Engine Air Inlets

This section deals with the engine air inlet (or air intake) as a system. The engine inlet is part of the aircraft and, while it does not comprise the engine, it nevertheless has (or can have) a profound effect on the performance of the engine. Furthermore, the position of variable geometry inlets used in many supersonic applications is (partly) dictated by the airflow requirements and general performance of the engine. This aspect of air inlets is covered in the following Section 6.2.

Engine inlets for traditional subsonic commercial aircraft applications are deceptively simple. While they appear to be little more than ducts of very short length, sometimes

Gas Turbine Propulsion Systems, First Edition. Bernie MacIsaac and Roy Langton.
© 2011 John Wiley & Sons, Ltd. Published 2011 by John Wiley & Sons, Ltd.

referred to as 'pitot'-type inlets, they have a number of subtleties that must be dealt with to ensure smooth operation under all flight conditions.

Military aircraft may require special considerations dependent upon the aircraft design and its mission. For applications where stealth is critical, for example, the air inlet to the engine may have to be arranged so that front face of engine is not directly visible to enemy radar from below or in front of the aircraft. In many cases, the engines of military aircraft are buried within the fuselage or wing root to provide a more optimum aerodynamic solution, particularly in supersonic applications.

In supersonic flight conditions, the inlet takes on more of the role of compression and expansion of the overall propulsion system. This implies the need for maximum overall system efficiency to which the inlet system is a contributor. Such a state of affairs usually dictates variable geometry and thus some form of control.

6.1.1 Basic Principles

The purpose of an engine inlet is to provide a duct through which outside air is presented to the engine inlet. The duct should ideally be as smooth and straight as possible in order to present a uniformly distributed airflow with minimum turbulence to the fan or compressor face. Furthermore, the axial velocity of the air at the compressor face is of the order of 500 ft/s (approximately Mach 0.5). This suggests that at flight speeds of Mach number less than 0.5 ($M < 0.5$), the duct will be subjected to an accelerating flow. At flight speeds of $M > 0.5$, however, the duct will be acting as a diffuser. In addition, the various attitudes that an airplane can achieve during flight suggests that the inlet duct will be required to turn the flow through an angle roughly equivalent to the aircraft angle of attack. Figure 6.1 shows the major features of the subsonic inlet system for a modern turbo fan engine.

As can be seen from the figure, the engine must draw in air from a larger area at low speeds ($M < 0.5$) and accelerate this flow to match the axial flow speed at the face of the engine.

The shape of the inlet duct is such that it has a minimum cross-sectional area at or near the inlet mouth. This forms a subsonic venturi with the free-streaming airflow at low subsonic speeds. At the higher speeds, the free stream area required to supply the engine is less than the inlet throat area; the inlet therefore acts as a diffuser to slow the air down to match the engine requirements. Excess air beyond that required by the engine is

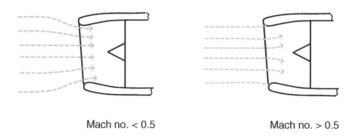

Mach no. < 0.5   Mach no. > 0.5

Figure 6.1   Typical subsonic inlets.

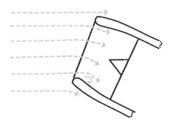

Figure 6.2   Inlet performance during climb.

spilled around the cowling which increases overall drag on the aircraft. This is a matter of concern to the designer of the aircraft, and the throat area of the duct must be set as a best compromise between aircraft drag and engine performance at flight speeds.

Some engine inlet designs incorporate spring loaded 'blow-in' (or 'suck-in') doors around the periphery of the nacelle to provide additional inlet area when the engine is operating at high rotational speed while the aircraft is flying at a relatively low speed (e.g., during takeoff and climb-out). This arrangement allows the aircraft designer to minimize the compromise mentioned above and to provide a more optimum inlet for the cruise condition.

At attitudes such as those which occur during takeoff and climb, the orientation of the inlet duct to the free stream can be at a substantial angle, as shown in Figure 6.2.

Under the conditions shown in the figure, there is a possibility of flow separation inside the bottom lip of the intake duct. Fortunately, the flow is still in an accelerating mode and therefore much less likely to separate. It will however generate increased turbulence and it is this condition that largely sets the throat area of the inlet.

Similar circumstances occur during takeoff in high crosswind situations prior to rotation and in side-slip during flight.

The possibility of ice formation and shedding of this ice into the engine is a major concern for aircraft propulsion system designers. To contend with this problem, the contoured outer lip of the cowling is usually fitted with some form of anti-ice protection. Cowl anti-icing systems are described in Section 6.3.

6.1.2   Turboprop Inlet Configurations

The description above serves as an introduction to the concepts of engine inlet design. Because the air inlet system is really part of the aircraft (or ship in a marine propulsion application), there are however a large number of other configurations possible. Air inlets associated with the turboprop engine are described below.

A turboprop engine delivers power to the aircraft through a propeller. A typical propeller operates at around 2000 rpm and absorbs about 1000–2000 HP in most common applications. The aerothermodynamics of turbine designs puts turbine speeds for this horsepower range at 20 000–30 000 rpm. This mismatch in speed requires a gearbox with a gear ratio of the order of 10 to 1. For front-mounted propellers, this arrangement requires the inlet ducting to obtain air from in front of the propeller and present it to the compressor downstream of the gearbox.

For epicyclic gearboxes, the configuration most commonly employed is an annular duct which is faired around the gearbox and which provides as smooth a presentation of air as is possible. Figure 6.3 is a schematic of this configuration. By contrast, gearboxes with a lay shaft tend to be offset from the centerline of the engine. This arrangement favors an air scoop typically on the underside of the gearbox; this arrangement is shown in Figure 6.4.

Finally, there is the ubiquitous Pratt & Whitney Canada PT-6 engine first mentioned in Chapter 1, which flies 'backwards'. This arrangement drives the propeller through a gearbox from the hot end of the engine. Air is ducted around the engine and enters from an annular/radial inlet at the back. The front portion of this arrangement is an air scoop on the underside of the engine; however, to avoid problems with icing, the inlet is oversized and some air is ejected out the back while only a portion enters the engine. The higher momentum associated with water droplets and/or ice crystals makes them reluctant to make the 90° turn to the engine; they are therefore ejected with the excess air as depicted in Figure 6.5.

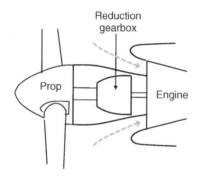

Figure 6.3   Turboprop inlet system for epicyclic gearbox drives.

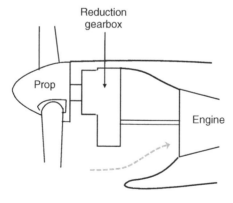

Figure 6.4   Turboprop inlet system for offset gearbox drives.

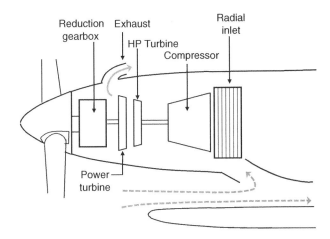

Figure 6.5   PT-6 inlet system arrangement.

6.1.3   Inlet Filtration Systems

For specific applications where solid particulates are commonly carried by the air, inlet filtration is required. These applications include helicopter operations in the presence of dust and sand, naval ships fitted with gas turbines, and, while less common, the adaptation of turboshaft engines to the propulsion of hovercraft.

Two types of separators are in common usage: the so-called barrier filters and the inertial separators. Depending on the conditions of operation, that is, the size and concentration of particulate found in the ambient air, one or both of these filters may be employed.

The barrier filter is little more than a fibrous material that is porous but capable of trapping particulates above a specific size. The media may be fiberglass or treated paper. It will likely be pleated to increase the surface area available for capture and retention of particulate, while at the same time keeping the frontal area constant.

Barrier filters are rated from the perspective of particle capture efficiency. Typical ratings suggest efficiencies of the order 95% for particulates of 2 μm and 99.5% for particulates of the order 10 μm. It is commonly noted that such filters tend to achieve higher efficiencies as they trap more material; however, this efficiency improvement is at the expense of increased pressure drop.

Pressure drops across well-designed barrier filters will be of the order of 1 inch of water at the start of their life. The filter media will either be washed and reused or discarded in favor of a new filter element as the pressure drop increases. Filters are commonly removed from service when the pressure drop exceeds 2.5–3 inches of water.

The inertial separator is nothing more than a large array of turning vanes or swirl vanes whose purpose is to impart angular momentum to the inlet airflow. This action forces any entrained particulate radially outward, allowing clean air to enter the compressor inlet.

Separation efficiency for inertial separators is determined by the mass of the particulates and the amount of swirl imparted to the flow. These filters are rather ineffective for particulate below 10 μm. Even at this particulate size, they are seldom better than 95%.

Their use is normally restricted to the removal of large particulates such as sand (in desert operations) or water droplets, where it is possible to get the water droplets to coalesce so that they can be collected and drained away by gravity.

6.2   Supersonic Engine Air Inlets

For all subsonic operations of the gas turbine engine, the inlet is well matched to the engine in so far as airflow is concerned. There may be more or less pressure drop accompanied by various levels of turbulence; however, the engine will be able to draw the amount of air required for operation with reasonably good efficiency at any power level.

As sonic speed is approached, however, the pitot type of inlet becomes less efficient as a shockwave begins to form at the inlet lip. At supersonic speeds this normal shock becomes so severe that the pitot type of air inlet becomes significantly inefficient; most of the forward speed of the aircraft is not recovered as pressure downstream of the shock but is heat dissipated at the shockwave itself.

In spite of the reduced efficiency of this concept, most of the earlier inlets for supersonic military aircraft employed fixed geometry. To compensate for the inherent inlet inefficiency, the additional required thrust (usually via afterburning) was provided. The F-100 Super Saber is a classic example of the simple pitot inlet approach used by many early supersonic fighter aircraft.

The F-16 Fighting Falcon produced by General Dynamics from the 1970s is a more recent example of a supersonic fighter aircraft that uses a fixed geometry pitot-type inlet (in this case, below the nose of the aircraft), as indicated in Figure 6.6. Trade studies

Figure 6.6   F-16 fighter showing the air inlet configuration (courtesy of USAF Photo, Sr. Airman Julianne Showalter, USAF Photo, Tech. Sgt. Michael Holzworth).

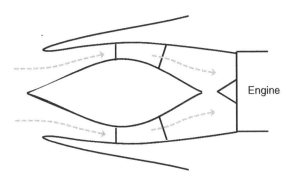

Figure 6.7   External/internal compression inlet concept.

into alternative inlet designs showed that the performance benefits of a variable geometry inlet were not sufficient to justify the added weight and complexity associated with its implementation. It should also be recognized that the benefits of a complex variable geometry inlet do not become significant until Mach numbers of about 1.3 and above; only aircraft with mission Mach number requirements substantially higher than this will benefit from the added weight and cost of its implementation.

An alternative approach used to improve inlet efficiency in supersonic flight, developed in the 1960s, is the external/internal compression inlet used on supersonic aircraft such as the British English Electric Lightning and many of the Russian Migs. This approach, depicted in Figure 6.7, is also a fixed geometry inlet. However, it produces a number of mild or 'oblique' shockwaves along the inlet as the air is slowed down before reaching the inlet throat, thus improving the pressure recovery and hence the operational efficiency of the inlet at supersonic speeds.

6.2.1 Oblique Shockwaves

To understand shockwave formation, consider a simple wedge-shaped body traveling through the air as indicated in Figure 6.8.

We know from basic fluid mechanics that there is a stagnation point at the nose of the projectile where the flow has stopped relative to the moving body. We know further that pressure waves travel in air at the speed of sound. Thus, under subsonic conditions, the projectile presents a pressure difference which is propagated upstream. The air in front of the projectile has time to react and create streamlines that will allow the flow to separate around the projectile in a smooth and continuous manner.

At the speed of sound (and all speeds above the sonic conditions), the projectile is traveling at a speed too great for the pressure waves to propagate upstream. Since the Mach number (in relative terms) is above unity upstream and must be zero at the leading edge of the projectile, and pressure waves cannot communicate the difference, a shockwave or discontinuity must occur somewhere in front of the projectile. On the axis of travel, this shockwave will be normal to the direction of travel. At points off the axes, the shockwave will bend away from normal and will form an oblique shock which, at points well away from the projectile, will become vanishingly small. As the speed is increased,

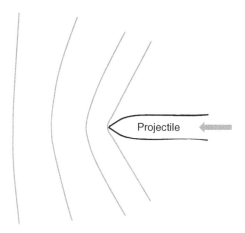

Figure 6.8 A projectile traveling in air.

the shockwave will attach itself to the leading edge of the projectile. A shockwave cone will form at some angle which is related to the speed of travel.

It is now known that a normal shockwave (i.e., a shock perpendicular to the direction of flow) will produce a sudden change in stagnation conditions across itself and that the flow downstream of the normal shock is always subsonic [1]. For example, a normal shock at Mach 2 will produce the following conditions:

$$\frac{P_{T2}}{P_{T1}} = 0.721$$

$$M_2 = 0.577$$

where P_{T1} is the total air pressure upstream of the normal shockwave; P_{T2} is the total air pressure downstream of the normal shockwave; and M_2 is the Mach number downstream of the normal shockwave.

In other words, a single normal shockwave will reduce the downstream flow to subsonic conditions ($M_2 = 0.577$) with a substantial pressure loss.

Let us now consider the possibility of using oblique shocks to improve the pressure recovery of the flow. An oblique shock will occur under conditions of a concave corner whereby the flow is bent through some angle as shown in Figure 6.9.

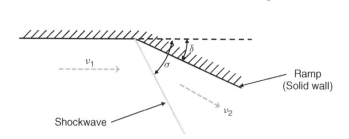

Figure 6.9 Flow through an oblique shockwave.

Under the conditions shown in the figure, the shockwave will stand at an angle σ relative to the upstream velocity. This angle σ will depend on the upstream Mach number and the angle δ that the solid wall ramp makes with the free stream flow.

Again, using an upstream Mach number of 2.0 and setting the ramp angle at $10°$, an oblique shock will stand at $40°$ relative to upstream conditions. In this case, the downstream conditions will be as follows:

$$\frac{P_{T2}}{P_{T1}} = 0.981$$
$$M_2 = 1.58.$$

Since the flow is still supersonic, it is feasible to repeat this process with a second ramp resulting in the formation of a second oblique shockwave. The conditions downstream of the second shock are as follows:

$$\frac{P_{T2}}{P_{T1}} = 0.986$$
$$M_2 = 1.26.$$

If we now arrange for a final shockwave normal to the flow, the air will exit the shock system subsonically with final flow conditions:

$$\frac{P_{T2}}{P_{T1}} = 0.987$$
$$M_2 = 0.807.$$

The above multi-shock process translates to an overall total pressure ratio across the three shocks of 0.955 compared to 0.721 for the single normal shock. A supersonic air inlet design that develops a number of oblique shockwaves prior to the final normal shockwave will therefore achieve a substantial improvement in inlet pressure recovery over the single normal shockwave inlet.

6.2.2 Combined Oblique/Normal Shock Pressure Recovery Systems

From the above example it is clear that an inlet with a number of oblique shockwaves prior to the normal shockwave would yield a much more efficient inlet design than the simple pitot type of inlet.

Let us now consider how we could harness these oblique shock systems to provide this more efficient inlet system for a jet engine in supersonic flight. The configuration shown in Figure 6.10 is one such approach, which indicates two oblique shocks both attached to the lower lip of the inlet cowl followed by a normal shock downstream of the nominal throat area. When the engine airflow demand does not match the inlet flow, the excess is spilled or bypassed around the engine. In many military applications, this excess flow is ejected along the upper surface of the wing.

A more sophisticated system involving three oblique shocks and the normal shock is shown in Figure 6.11. In this case, the front oblique shock is not attached to the lower lip, whereas the second and third shocks are attached. Such a system is sometimes referred

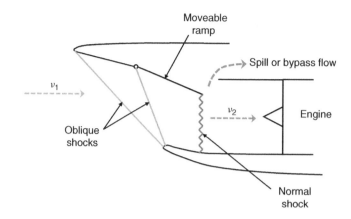

Figure 6.10 An oblique shockwave inlet for supersonic conditions.

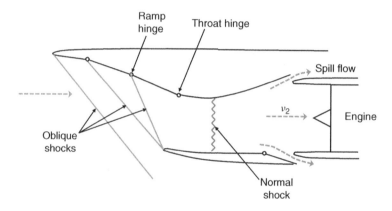

Figure 6.11 A mixed compression inlet with three oblique shocks and one normal shock.

to as a mixed compression inlet. This inlet arrangement has two hinges, the ramp hinge and the throat hinge, both of which are controlled and positioned in order to optimize the performance of the inlet. The compression is started by a weak oblique shock which is attached to the leading edge of the inlet. The compression is continued through weak oblique shocks to the minimum area or throat section, after which a normal shock reduces the velocity of the flow to subsonic conditions.

From there, the flow is further diffused to a subsonic speed which matches that required at the engine compressor face. Several aspects of this arrangement are worthy of discussion.

The location of the normal shock is controlled by the back pressure in the inlet duct downstream of the normal shockwave. If this pressure is not correct, the shock will move. Too high a back pressure will force the normal shock further upstream until it reaches the throat, after which it will be disgorged and the inlet is said to have 'unstarted'. Too low a back pressure will move the shock in the direction of the engine. This condition

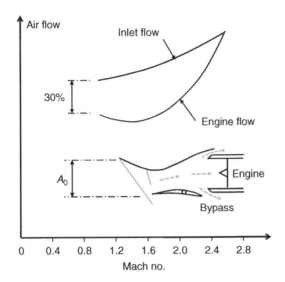

Figure 6.12   Airflow mismatch of a turbofan engine and a typical 2D inlet.

suggests that too much of the pressure recovery is being lost across the normal shock and the overall efficiency of the inlet will be lowered.

The orientation of the oblique shocks is entirely a function of the local Mach number upstream of the shock and the turning angle imposed by the shape of the wall which is, of course, determined by the angle established by the movable ramps. Ultimately, the local Mach number is determined by a combination of aircraft flight Mach number and the airflow through the inlet. At some operating point, the airflow through the inlet is a good match for the engine; however, for all other points in the flight regime, this situation does not exist and any air in excess of that required by the engine must be spilled or bypassed in some way. This is illustrated in Figure 6.12 for a typical two-dimensional inlet.

In the example shown in the figure, the match point between the inlet flow rate for control of the shock system and that required by the engine occurs at a Mach number of about 2.5. At operating points between $M = 1.0$ and $M = 2.5$, as much as 35% of the airflow must be bypassed.

Finally, in the region of the throat and downstream from it (i.e., in the region of the position of the normal shock), some form of boundary layer bleed control is required. This is due to the messy system of shock/compression waves that exist as a consequence of the normal shock meeting the boundary layer at the inlet wall. In this region, the wall is usually made porous and the amount of boundary layer bleed air may or may not be controlled. It is undeniably better to control the amount of boundary layer bleed; however, this is a matter of cost which needs to be justified on the basis of the improvement in the overall propulsion system performance that can be achieved.

6.2.3   Supersonic Inlet Control

The above description indicates that a sophisticated multidimensional control system, whereby the inlet geometry is controlled by measurements of local flow conditions set

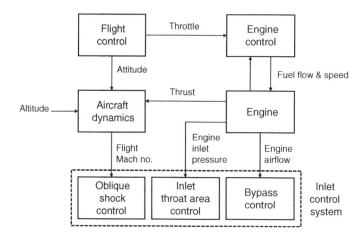

Figure 6.13   Overall block diagram of the aircraft, inlet, and engine dynamics.

up by the flight regime of the aircraft and optimized around finding a match between the pressure recovery achieved by the shock system and the airflow requirements of the engine, is desirable. A top-level overall block diagram of the system, as seen from the aircraft designer's perspective, is shown in Figure 6.13.

Each of the individual elements of the inlet control system (oblique shock control, inlet throat area control, and bypass control) will be described in the following sections.

6.2.3.1   Oblique Shock Control

The angles at which the shock system is set are controlled via a simultaneous positioning of ramp angle, inlet throat area, and bypass door position. Reference to Figure 6.11 is helpful in visualizing this system of geometry control.

The inlet ramp angle is usually scheduled as a function of local Mach number which can be inferred from measurements of static and total pressures obtained near the ramp. The actuator is commonly a hydraulic ram or an electro-hydraulic rotary actuator. A block diagram of the ramp angle control is shown in Figure 6.14, indicating how ramp angle is controlled in an open-loop fashion as a function of local Mach number.

6.2.3.2   Throat Area Control

The inlet throat area is also controlled open loop through a schedule which is a function of the local Mach number. Boundary layer bleed is usually associated with the inlet

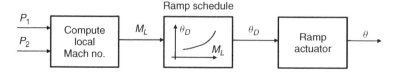

Figure 6.14   Inlet ramp angle system diagram.

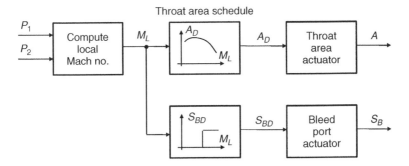

Figure 6.15   Throat area control system diagram.

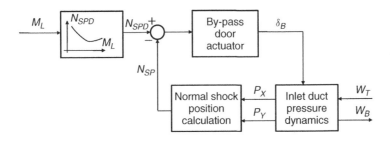

Figure 6.16   Bypass door control system diagram.

throat area control. The wall is made porous in this region and the air bleed is controlled through discharge ports which are two-position only: closed for subsonic flight and open for supersonic flight. A system diagram for the throat area control is shown in Figure 6.15.

6.2.3.3   Bypass Flow Control

Finally, the bypass door is controlled closed loop on pressure measurements which infer the location of the normal shock in the area of the throat. This control establishes a desirable position of the normal shockwave and compares it to the 'measured' position and makes adjustments accordingly. A system block diagram is shown in Figure 6.16.

6.2.4   Overall System Development and Operation

The foregoing description indicates that control of the inlet system is partly open loop, suggesting that a substantial amount of development testing will be required to refine the schedules involved and to match the inlet to the engine under all flight conditions.

During engine start-up, the inlet will be operating in the subsonic regime. Under such conditions, the inlet ramps and throat area are opened as wide as possible and the bypass door is folded inward to maximize the area available for subsonic flow to the engine. In other words, external flow is sucked in through the bypass door area by the engine during this regime of flight.

During takeoff conditions, the throttle of the engine will be at a maximum during which time the airflow demand is at its highest. However, flight is still subsonic. Under such conditions, ramps are retracted such that the inlet throat area is at a maximum. Similarly, the bypass door is still folded inward. As aircraft speed is increased, the effect of ram recovery is increased and the bypass door is closed progressively.

In the transition from subsonic to supersonic flight, the shock systems will be forming and will, in any event, be external to the inlet. This is a high drag condition which should be transitioned as quickly as possible.

At about a Mach number of 1.3–1.4, the shock system will likely begin to attach to the aircraft and move progressively into the inlet. At this point, the ramps will begin to move and the throat area will assume a position in order to control the location of the shocks. In this flight regime, the bypass airflow will be at or near its maximum and the bypass door will be fully open to control the location of the normal shock. Above a Mach number of 1.5, the flight regime is usually considered to be normal and control of the geometry will be as per the schedules defined previously.

During the development stage, considerable emphasis will be placed on inlet performance under conditions of rapid flight maneuvers (dive, roll, yaw, etc.) during which the engine will be expected to cope with highly distorted inlet flow conditions. Should the engine stall due to excessive flow distortion during these maneuvers, its control will move rapidly to a low-throttle condition to avoid damage to the engine. This function is imbedded as a protective feature in the engine control and is completely independent of the aircraft and/or inlet control system. This action will completely overwhelm the inlet control since the engine airflow requirements will be cut sharply. It may therefore be expected that the duct pressure will rise very sharply and the entire shock system will be disgorged. Such events have been known to damage the inlet system and considerable effort is frequently expended on flow design details which soften or eliminate such pressure pulsations. In the event that the engine is only partially stalled (rotating stall), the pressure fluctuations are smaller but more regularized. Under these conditions the shock system, especially the normal shock, will be destabilized resulting in a condition known as 'inlet buzz'. These conditions are outside the realm of systems control and are the rightful purview of the aircraft designer.

The final flight condition worthy of note is that of in-flight shutdown of the engine. Under these conditions, the ramps and throat area are moved to their minimum position and the bypass door is fully opened. This arrangement moves the shocks to a 'best position' intended to minimize drag on the aircraft.

6.2.5 Concorde Air Inlet Control System (AICS) Example

Perhaps the best example of a supersonic air inlet design is that of the Concorde supersonic transport. As shown in Figure 6.17, this aircraft has four engines located in two separate assemblies below and aft of each wing. Each engine has its own two-dimensional inlet similar to the concepts covered above.

For commercial viability, the overall propulsion system was required to operate with extremely high efficiency at the cruise flight conditions of Mach 2.0 at 50 000–60 000 ft. At the 60 000 ft top-of-descent condition the air inlet contributes most of the propulsion

Figure 6.17   The Concorde supersonic transport (courtesy of Airbus).

system thrust and develops an impressive overall compression ratio including the inlet and engine LP (low pressure), and HP (high pressure) compressors of about 80:1.

6.2.5.1   Air Inlet General Description

The Concorde AICS is described as a two-stream external compression design where the air entering the inlet is divided into primary and secondary flows. The primary flow enters the engine via the LP compressor while the secondary air passes through secondary air doors (SADs) around the engine, over the top of the jet pipe and primary exhaust nozzle, exiting the power plant via the secondary exhaust nozzle. Figure 6.18 shows the basic inlet arrangement.

As shown, the inlet is a two-dimensional ramp type with a variable area throat controlled by two movable ramps. The alternative supersonic inlet concept used by the SR-71 comprises a variable center-body (or 'shock cone') also known as an axisymmetric inlet. This type of inlet is actually more efficient at high Mach numbers than the two-dimensional type; however, its sensitivity to even moderate angles of incidence or side slip make this type of inlet unsuitable for commercial applications.

The Concorde inlet is reasonably insensitive to worst-case levels of side slip that could result from aerodynamic disturbances in supersonic flight. This was achieved by the incorporation of subtle modifications to the inlet cowl established during the flight test phase of the program. Side-slip correction algorithms included in the development AICS controller were found unnecessary and deleted in the production design.

A splitter plate is located between the two engines in order to give a degree of air flow independence and to prevent a shock expulsion event occurring in one engine from impacting its neighbor.

Above each pair of engines is a common fixed geometry diverter arrangement whose function is to limit wing boundary layer ingestion. Boundary layer growth within the inlet

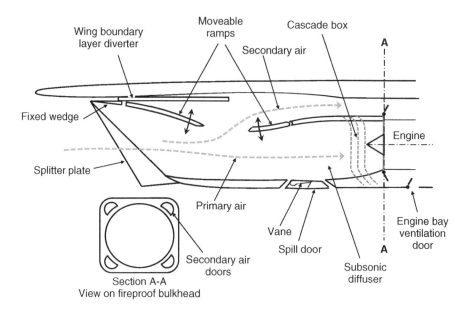

Figure 6.18   Concorde air inlet arrangement (courtesy of BAE Systems).

at very high local Mach numbers is also addressed by providing a series of holes located in the inlet floor (not shown in the figure). This feature was found necessary only on the inboard inlets of each pair.

The forward and aft movable ramps are linked together by torque tubes and levers. Four screw-jack actuators operate in unison to position the ramps and redundant hydraulic motors provide the rotary power.

Primary air passes through the inlet throat into the subsonic diffuser to deliver air at about Mach 0.5 to the compressor inlet.

Secondary air passes above the variable ramps to the secondary air doors (SADs) directly to the upper doors and via a cascade box to the lower doors. From there, secondary air passes through the engine bay and over the jet pipe before exiting at the variable area exhaust nozzle. The SADs are also driven by screw jacks driven by a flexible drive from hydraulic motors located at the lower section of the inlet.

A spill door located at the base of the inlet operates in conjunction with the ramp system to control the position of the normal shock during supersonic cruise, in order to accommodate changes in air inlet conditions and engine mass flow demand. A free-floating vane located within the spill door opens under aerodynamic force developed from engine airflow demand when the aircraft is operating below about Mach 0.7 to provide supplementary airflow to the engine. Above this point, the vane will remain fully closed. The spill door is operated by a hydraulic ram actuator with a redundant source of hydraulic power.

6.2.5.2   Basic Operation

The basic functional requirements of the air inlet system are as follows:

- to ensure that the airflow velocity reaches the engine LP compressor inlet at Mach 0.5;
- to modify the inlet capture area such that the airflow demanded by the engine is satisfied, ideally without spillage; and
- to maintain the inlet shock system in the 'critical' condition for maximum inlet efficiency.

The operating regions of the inlet can be focused on three primary areas of operation: the subsonic regime from static to Mach 0.7; the transonic regime from Mach 0.7 to 1.3; and the supersonic regime from Mach 1.3 to 2.0. The main aspects of inlet operation for each of these regimes are described below in sub-sections 'Subsonic Regime', 'Transonic Regime', and 'Supersonic Regime'.

Subsonic Regime

During takeoff and low-speed flight, the ramps are raised to the position of maximum capture area and the spill door is closed. In addition, the auxiliary vane will be open to provide a larger engine inlet capture area. This vane will move toward the closed position as aircraft speed increases and will become fully closed at about Mach 0.9.

Below about Mach 0.3 there is no secondary airflow and the SADs remain closed to prevent exhaust gases from entering the engine inlet via the inlet throat. Above this speed, SADs are opened and secondary airflow continues to increase.

Above Mach 0.55 the divergent section of the engine exhaust nozzle begins to open as a function of flight Mach number in order to effectively accommodate the increasing secondary airflow. This controls engine efflux divergence and limits flaring of the efflux. This feature of secondary airflow release is unique in two-dimensional inlet designs and provides both minimal drag and a small contribution to propulsion system thrust.

Most military supersonic aircraft with two-dimensional inlets vent secondary airflow to the atmosphere either alongside the inlet or above the wing. The drag associated with this approach, while not critical for military missions, is unacceptable for Concorde which must spend 2–3 hours at the supersonic cruise condition.

Transonic Regime

As the aircraft accelerates beyond Mach 1.0, the shock system begins to form. By Mach 1.3 it is fully formed with the fourth oblique shock attached to the lower intake lip (see Figure 6.19 which shows the complete shockwave system for the inlet for speeds above Mach 1.3).

During the transonic acceleration, the ramps are controlled as a function of LP compressor speed (N_1 rpm) to ensure correct intake mass flow matching to engine mass flow demand. The ramps will move progressively down and, if necessary, the spill door will open as N_1 decreases.

Supersonic Regime

As the aircraft accelerates to the design cruise speed, the second, third, and fourth shocks coalesce onto the lower lip of the inlet. The inlet is now controlled to maintain the normal shock at the optimum 'critical' position irrespective of changes in aircraft speed, ambient conditions, or engine mass airflow demand.

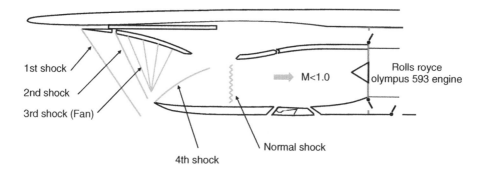

Figure 6.19   Concorde inlet shockwave system (courtesy of BAE Systems).

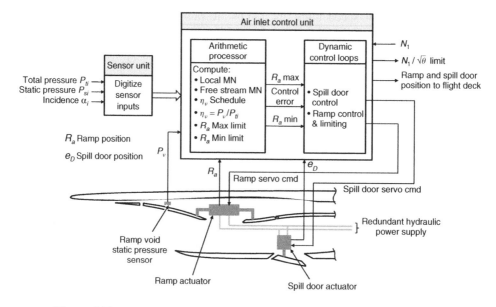

Figure 6.20   Concorde air inlet control schematic (courtesy of BAE Systems).

This is achieved by indirect measurement of pressure recovery using the ramp void static pressure relative to the free stream total pressure as the scheduled parameter.

A control system schematic diagram is shown in Figure 6.20, which illustrates the basic control strategy employed.

As shown, aircraft air data from the pitot-static and incidence sensors are digitized and fed to the air inlet control unit (AICU). Within the AICU, the arithmetic processor section computes local and free stream Mach numbers and the required η_v schedule for optimum inlet operation. This is compared to the measured η_v to obtain a control error which is input to the dynamic control loops section for control of the ramps and spill door actuators. Ramp maximum and minimum limits are also computed as a function of the prevailing incidence and flight condition.

For any finite value of control error, the ramps will continue to move at a rate proportional to the magnitude of the error (i.e., integral control action). If the ramps reach the prevailing maximum ramp angle $\delta_2\,Max$, before achieving the desired η_v, the spill door will be opened to satisfy the requirement and bring the control error to zero. Conversely, if the minimum ramp angle is reached, the engine speed N_1, and therefore mass airflow demand, must be reduced to satisfy the η_v requirement and maintain the normal shock in the optimum (critical) position.

In addition to the integral control action described above, rate-of-change of N_1 (i.e., N_1-dot) is used as a phase advance term in order to effectively accommodate rapid changes in engine speed.

This closed-loop control system arrangement, with the N_1-dot phase advance term, provides a very robust system allowing relatively free engine handling giving the ability to slam throttles from maximum continuous to flight idle and vice versa. It also yields an inlet performance efficiency of between 94 and 97% in the supersonic cruise condition, which contributes substantially to the Concorde's operational viability.

At the end of the cruise phase, the ramps will go down to their maximum limit as throttles are retarded and engine mass airflow is reduced. The spill door will also open to dump any excess airflow.

As flight Mach number reduces, the spill door will close leaving the ramps to control the inlet. By the time Mach number has reduced below 1.3, the ramps will be fully raised having completed their job for the flight.

6.2.5.3 The Division of Power-plant Thrust

It is interesting to learn how the thrust developed by the total power plant is divided up for the two operating extremes, that is, the slow flight/subsonic condition and the supersonic cruise condition. This is illustrated by Figure 6.21. As indicated, the differences

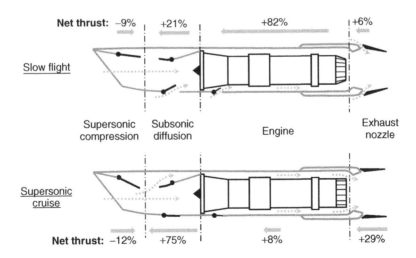

Figure 6.21 Concorde power-plant division of thrust (courtesy of BAE Systems).

contributed by the main components of the power plant between these two conditions are quite remarkable.

Consider initially the slow flight/subsonic condition. Here the thrust allocations are what we would expect with the engine contributing more than 80% of the total. The inlet at this flight condition generates a net positive thrust of 12% from the pressure recovered in the diffuser section of the inlet. A positive 6% thrust is also generated by the exhaust nozzle.

This picture changes dramatically in the supersonic cruise condition. In the cruise phase, the air inlet generates the majority of thrust while the engine itself contributes only 8%. The exhaust nozzle now contributes almost 30% of the net thrust. Note that the contribution of the supersonic compression section of the inlet is negative and is therefore a drag term and the total net thrust in each case adds to 100%.

It should also be noted that while the engine makes only a minor contribution to the overall power-plant thrust at cruise, without the engine the system falls apart and the various inlet and exhaust components would contribute precisely zero thrust.

The above example illustrates the importance of the air inlet (and exhaust) system designs in supersonic aircraft applications.

6.3   Inlet Anti-icing

The air inlet to the engine must be protected against the formation of ice; this can occur when operating in environmental conditions that support ice accretion on the inlet cowl or on the fan/compressor inlet guide vanes or stators. During flight through clouds at medium to low altitudes, super-cooled water droplets coming into contact with aircraft components at stagnation locations (i.e., where air flow is stationary) such as the inlet cowl leading edge or the first- or second-stage compressor stator vane leading edges, can form and accumulate ice. In ideal icing conditions, the ice accretion can occur very quickly with the potential for airflow restriction and/or distortion leading to major loss of performance or even engine shutdown. There is also the potential for ice becoming dislodged, resulting in damage to the engine from ingestion into the compressor.

The aircraft wing leading edge is subject to the same problem where ice accumulation can seriously impact the aerodynamic performance of the wing with potentially catastrophic consequences.

An anti-icing system is therefore required to effectively prevent ice formation in these critical stagnation areas to ensure safe operation in flight conditions that support ice formation on the aircraft or engine structure. Clearly such an anti-icing system must be reliable, easy to maintain, and involve a minimum weight penalty.

There are two methods in common use today for engine anti-icing:

1. the use of hot air from the compressor; and
2. electric heating of specific areas.

In some applications both electrical heating and hot air methodologies are used; however, the electrical heating technique is historically more common in turboprop applications.

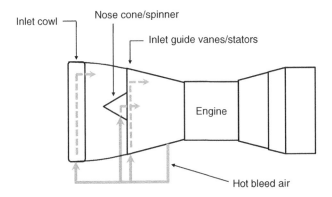

Figure 6.22   Bleed-air anti-icing system overview.

6.3.1   Bleed-air Anti-icing Systems

Bleed-air anti-icing systems use high-temperature air from one of the higher stages of compression typically shared by other bleed-air system needs such as cabin pressure regulation and environmental control. Generation and control of the bleed-air source is described in Section 8.2.

High-temperature bleed air is ducted to the inlet cowl, the nose cone (or the spinner in the case of the turboprop application), and through internal passages within the fan and compressor inlet guide vanes. In some applications, the first stage compressor stators also receive bleed air for ice protection. Bleed air is discharged into the engine inlet to ensure a continuous flow of air though the anti-icing ducting (see Figure 6.22).

6.3.2   Electrical Anti-icing Systems

Electrical anti-icing systems are in common use in turboprop applications to prevent ice accumulation on the propeller leading edge as well as the air inlet cowl and spinner. In today's modern composite nacelles, electrical heating coils are sandwiched within the carbon-fiber layers. Two separate circuits are typically used to provide both continuous and selectable heating.

When anti-icing is selected, the selectable heating elements can be energized using different duty-cycles (i.e., the ratio between the on time and off time). Two levels of anti-icing protection are typically available to provide appropriate protection against either moderate or severe icing conditions.

6.4   Exhaust Systems

Chapter 2 summarized the principles of the simple jet pipe propelling nozzle of a turbojet engine, describing how the thrust generated by the engine is a combination of the nozzle exit plane pressure relative to the local ambient pressure acting on the nozzle area and the rate of change in the momentum of the airflow through the engine.

Chapter 4 covered afterburner control and how it is critical to match the exhaust nozzle area to the afterburner thrust setting in order to maintain a consistent compressor operating point, as nozzle pressure varies with afterburner fuel flow.

In this section, we will address additional exhaust system concepts and the performance benefits they bring to the gas turbine thrust engine propulsion system, including:

- thrust reversing systems; and
- thrust vectoring systems including V/STOL applications.

6.4.1   Thrust Reversing Systems

Thrust reversing is a standard feature of all of today's modern jet-powered commercial aircraft and is also in common use in military applications. As described in Chapter 5, turboprop applications use negative propeller pitch to provide reverse thrust during landing and during flight in some military applications, where a high rate of descent is a key tactical requirement.

The thrust reversing systems described here focus on the thrust generating gas turbine engine, namely the turbojet and turbofan applications.

In commercial aircraft applications, the early jet-engine-powered aircraft created a need for reduced landing distance capability over and above the use of traditional wheel braking systems. This need was a result of the higher landing speeds typical of jet-powered aircraft, together with existing runways which were often inadequate in length for consistent and safe operation.

The approach taken in providing thrust reversal is to reverse the direction of the engine exhaust gas stream and use the power of the engine itself to generate a deceleration force. While it was considered impractical to completely reverse the thrust of the engine, by diverting the efflux through an angle of about 45° a reverse thrust of about 50% of maximum forward thrust can be accomplished. As a result, the landing distance of a typical commercial jet-powered aircraft can be reduced by 20–25% while reducing the severity of wheel brake usage with attendant savings in brake maintenance costs. Thrust reversing is also much more effective and safer than using only wheel brakes on wet, icy, or snow-covered runways.

Two of the most common types of reverser design in service today are described below.

- The clamshell door concept is used in turbojet and mixed flow bypass engines, and is illustrated in the conceptual schematic of Figure 6.23.
- The cold stream reverser with hot stream spoiler is used in the high bypass ratio turbofan engine and involves deflection of both the fan and core exhaust streams. Figure 6.24 illustrates this thrust reversing concept.

The latter reverser design is based on the fact that most of the engine thrust from high-bypass turbofans comes from the fan rather than the core and therefore it is sufficient to just 'spoil' the core exhaust to eliminate any forward thrust component. In some applications, however, a more efficient core deflection arrangement is used by deflecting the core exhaust gases forward through cascade vanes to provide some positive reverse thrust. This component is typically less than 10% of the total reverse thrust, so the tradeoff in terms of additional weight and complexity may not be justified.

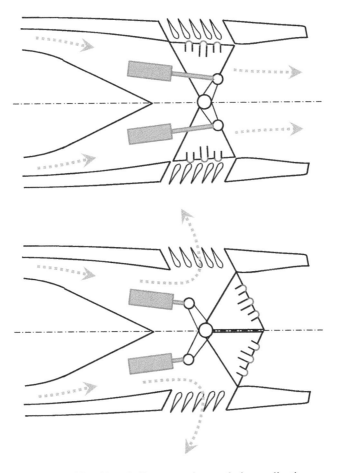

Figure 6.23   Clamshell reverser in a turbojet application.

6.4.1.1   Thrust Reverser Actuation and Control

The functional integrity of the thrust reverser system is critical since inadvertent deployment in flight (particularly at high speed) can be catastrophic. A number of safety-interlocks must therefore be provided to ensure that the system remains in the stowed position until thrust reversal is selected by the flight crew during the landing process. Specifically, the following safeguards are typically provided within the selection and actuation system:

- reverser deployment is inhibited until the throttle lever has been retarded to the idle detent;
- a mechanical locking device prevents thrust reverser operation if an actuation system fault is detected, for example, actuation system power supply failure; and
- once the thrust reverser system deployment is initiated, the throttle lever cannot be advanced until the actuating mechanism is at or near the fully deployed position.

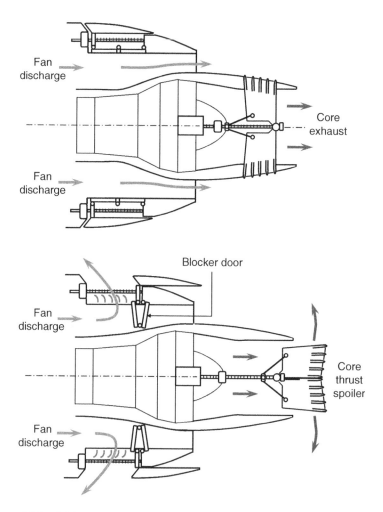

Figure 6.24 Cold stream reverser and hot stream spoiler for a turbofan application.

Both pneumatic and hydraulic actuation power supply techniques are in common use in today's thrust reversing systems. In the case of pneumatics, the high pressure air source is provided by engine bleed air. In the case of hydraulic power, a dedicated oil pressure source which uses engine lubrication oil as the fluid source together with a dedicated engine-driven gear pump (with an appropriate over-pressure relief valve) provides motive power to the various actuators. Alternatively, the thrust reverser actuators may be powered by the aircraft hydraulic system or high pressure bleed air.

Figure 6.25 shows one example of a thrust reverser actuating system using high pressure bleed air as the motive power source. This schematic, which shows the actuation of fan discharge blocker doors and a core thrust spoiler typical of a high-bypass turbofan, should be viewed in conjunction with Figure 6.24.

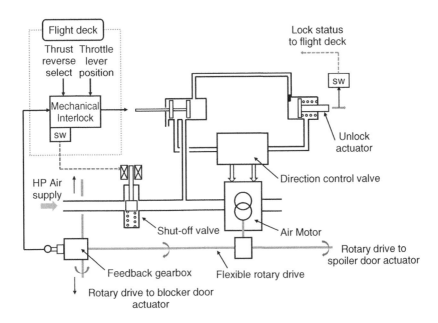

Figure 6.25   Trust reverser schematic.

As shown in Figure 6.25, high pressure air passes through a shut-off valve which is actuated by the selection of reverse thrust by the flight crew. From the shut-off valve, air is fed to an air motor and a spool valve. The latter, positioned by the throttle linkage, determines the rotational direction of the air motor. This same spool valve directs air to the lock/unlock valve, allowing deployment of the thrust spoiler and blocker door screw-jacks which are in turn actuated via a flexible rotary drive powered by the air motor.

A feedback gearbox in the flexible drive circuit provides a mechanical input to a mechanical interlock arrangement, which prevents the crew from increasing engine thrust until the thrust reversing mechanism is close to fully deployed.

If there is a failure of the air supply, the lock/unlock valve will remain in the locked position and the thrust reverser system function will be inhibited.

6.4.2   Thrust Vectoring Concepts

In addition to thrust reversing there are two other aspects of thrust vectoring that are in common use today, specifically:

- aircraft that use jet thrust to provide vertical takeoff and landing (VTOL) capability; and
- aircraft that use in-flight thrust vectoring to enhance maneuverability and agility.

The first public demonstration of VTOL took place at the Farnborough Air Show in the mid 1950s when the experimental craft dubbed 'the flying bedstead' showed its ability to takeoff, maneuver, and land under jet power.

Soon after, the Pegasus lift/propulsion engine was developed by Bristol Engines (later to be acquired by Rolls-Royce). This engine is essentially a turbojet with a two-stage free turbine driving a large front fan. Two separate exhaust nozzles located on each side of the engine, one using fan discharge air and the other using engine exhaust, have the ability to deflect the efflux from the normal aft direction to just forward of fully vertical.

A cutaway of the Pegasus is shown in Figure 6.26. See also [2] for a more detailed discussion of VSTOL concepts.

Varying the thrust direction is achieved by simply rotating a series of slat-type guide vanes at the outlet of each nozzle exhaust. Another important feature of the engine design is that the resultant thrust vector remains in a fairly consistent location relative to the engine's center of gravity over the full range of throttle and nozzle deflection settings.

The Pegasus engine is used to power the Harrier aircraft which is still in service with the Royal Navy. The US Marines use the same basic concept in their AV-8B Harrier aircraft which was built under license by McDonnell–Douglas (now Boeing) and is still in service today. A general arrangement of the Harrier aircraft is shown in Figure 6.27.

The plan view clearly shows the over-sized fan section of the engine. The dark arrows show the location of the vertical thrust vectors when in the hover mode. The lighter arrows indicate 'puffer jets' which use fan discharge air to provide pitch and roll stabilization, since the conventional flight control surfaces are totally ineffective in the hover and at low forward speeds.

While the Harrier is capable of vertical takeoff, hovering, and landing, in service the practical mode of operation is referred to as short takeoff and vertical landing (STOVL). This is much more efficient in terms of fuel usage. Navy aircraft are fitted with specially designed ramps called 'ski-jumps' to facilitate effective short takeoff capability.

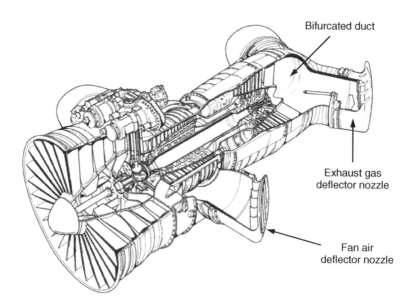

Figure 6.26   The Rolls-Royce Pegasus lift/propulsion engine (© Rolls-Royce plc 2011).

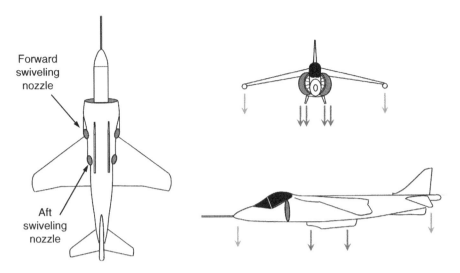

Figure 6.27   General arrangement of the BAE Harrier.

It is interesting to note that during the war between Great Britain and Argentina in the 1980s, Harrier aircraft pilots developed a new dog-fighting technique called vectoring in flight or VIF-ing. If a Harrier found itself with an enemy aircraft on its tail, the pilot would select full down vector momentarily. As a result, the enemy aircraft had no choice but to overshoot the Harrier only to find itself now in an unfavorable location ahead of the Harrier.

A more modern version of the VTOL concept is implemented in the US Marines version of the new Joint Strike Fighter (The F-35B Lightning). This aircraft employs a significantly different approach from the Harrier as indicated by the conceptual drawing of Figure 6.28.

As shown, there is a separate lift fan located to the rear of the cockpit driven mechanically through a clutch by the F-135 engine's LP shaft, which protrudes forward of the engine inlet. The exhaust of the F-135 passes through an articulating nozzle which can be rotated between normal axial thrust and the vertical position shown in the figure. Fan/LP compressor discharge air from the F-135 is used to provide roll control via roll post nozzles located outboard under each wing.

The F-135 turbofan engine (also classified as a 'leaky turbojet' with a bypass ratio of less than 1.0) comprises a three-stage fan/LP compressor and a six-stage HP compressor. The maximum thrust of the F-135 in normal forward flight is 28 000 lb (dry), increasing to 43 000 lb with afterburner.

In the hover mode, the F-135 engine transitions to approximately 50% turboshaft and 50% turbojet. The LP shaft driving the lift fan delivers up to 35 000 SHP in this mode, which converts to 20 000 lb of lift. The jet exhaust contributes up to 18 000 lb of direct vertical lift and the roll post jets use LP compressor discharge air to deliver up to 1950 lb of thrust each.

The US Air Force F-22 Raptor uses thrust vectoring primarily to provide improved agility and maneuvering in flight by the provision of thrust vectoring in the pitch axis; a nominal range of $\pm 20°$ from the normal thrust axis of the aircraft is available.

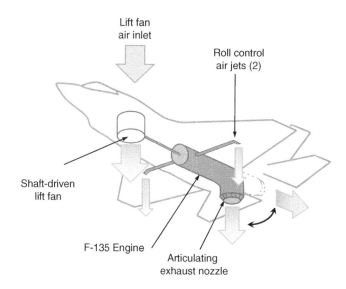

Lift fan
air inlet

Roll control
air jets (2)

Shaft-driven
lift fan

F-135 Engine

Articulating
exhaust nozzle

Figure 6.28 Conceptual drawing of the F-35 in Hover.

Figure 6.29 shows a sequenced photograph of the F-119 Engine under test at the Pratt & Whitney test facility in West Palm Beach, Florida, showing the full range of thrust vectoring available in the pitch axis. This capability allows the pilot to impart a significant pitching moment to the aircraft which can be particularly important at low speeds where the aerodynamic control surfaces become much less effective. To accomplish this a complex nozzle exhaust actuating mechanism is employed which, as indicated in the figure, is quite large.

The geometric principle employed by this type of mechanism, which provides both thrust vectoring and thrust reversing in an integrated actuation package [3], is shown conceptually in Figure 6.30.

The F-119 nozzle is described as a two-dimensional convergent-divergent (2D-CD) nozzle.

6.4.2.1 System Considerations

The primary system considerations with all of the above vectoring systems summarized above (VTOL, VSTOL, STOVL, and in-flight vectoring) are safety and functional integrity. This applies not only to the engine(s) but also to the actuation systems and to their associated power supplies. Single-engine aircraft such as the Harrier and F-35 depend entirely on the functional integrity of the main engine, particularly when operating in or near to the hover condition. It is therefore clearly important to minimize the exposure to operations in this critical mode of flight.

The thrust vectoring system for the F-22 should be considered as an extension of the flight control system, since it can generate significant pitching moments on the aircraft.

Figure 6.29   F-119 engine undergoing thrust vectoring tests (courtesy of Pratt & Whitney).

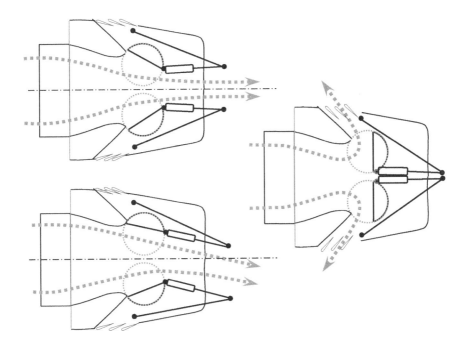

Figure 6.30   F-119 thrust vectoring and reversing nozzle concept.

The functional integrity of the nozzle actuation and control system must also be considered flight critical such that no single failure (or second failure following a first failure that remains dormant and undetected) shall cause loss of control. Failsafe considerations dictate that any failure of the vectoring system results in reversion to conventional axial thrust.

References

1. Shapiro, A.H. (1953) *The Dynamics and Thermodynamics of Compressible Fluid Flow*. Ronald Press, New York.
2. Rolls-Royce (1973) *The Jet Engine*, Rolls-Royce Ltd, (Part 21).
3. Treager, I.E. (1995) *Aircraft Gas Turbine Technology (Part 2 Section 8)*, 3rd edn, Glencoe/McGraw Hill.

7

Lubrication Systems

This chapter deals with the lubrication systems commonly used in modern jet engines. The basic principles of bearing and bearing lubrication are reviewed with the intention of providing technical background for the design and the selection of the lubrication system and its components.

The task of the lubrication system is to reduce friction and wear of moving and non-moving parts of the engine and to provide cooling as required by the bearing systems. These issues will be examined in this chapter in the context of the requirements imposed on the lubricant to provide these functions.

Finally, the design requirements of the lubrication systems will be considered in the context of commonly used system architectures. Each of the major components will be described together with system issues of location and health monitoring methods.

7.1 Basic Principles

The requirements for lubrication systems are derived from the need to reduce friction and wear at the interface between moving parts of a machine.

Consider the problem of sliding friction as presented in Figure 7.1. In this diagram, the upper body is moving relative to the lower body and is supporting a load of magnitude N. Application of the basic laws of physics allows us to estimate the friction force that must be overcome to move the block.

This is written as:

$$F_f = \mu N \tag{7.1}$$

where μ is the coefficient of friction between the two surfaces and F_f is the friction force parallel to but opposite to the direction of motion.

The coefficient of friction is an empirically derived parameter which is related to the properties of the materials. These properties include the electromagnetic forces between the molecules which make up the materials in contact as well as the surface finish of the objects themselves. Considering dry static friction, the coefficient of friction between steel and steel of similar finish is approximately 0.8 whereas the coefficient of friction

Gas Turbine Propulsion Systems, First Edition. Bernie MacIsaac and Roy Langton.
© 2011 John Wiley & Sons, Ltd. Published 2011 by John Wiley & Sons, Ltd.

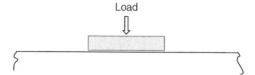

Figure 7.1   Surfaces in relative motion.

between steel and Teflon is only about 0.04. This difference is primarily due to the lack
of interaction at the molecular level between the two materials.

The addition of a fluid capable of affecting the separation of the two surfaces has a
dramatic effect on the coefficient of friction. In this instance, the shearing of the liquid
layer determines the apparent coefficient of friction, which is almost entirely related to
the viscosity of the liquid. The overall frictional force is therefore greatly reduced.

The concept of using two smooth surfaces separated by a film of oil is the essential
element of hydrodynamic bearings. Such a bearing arrangement is shown in Figure 7.2 [1].

In Figure 7.2a, the shaft is stopped and the journal is in direct contact with the bearing.
In Figure 7.2b, the shaft has just begun to rotate in the clockwise direction and oil is
introduced at the top of the bearing. The clearance between shaft and bearing are such
that the oil will flow around the shaft, lifting the shaft upwards and to the left. As steady-
state conditions are reached as depicted in Figure 7.2c, the shaft is seen to be riding on a
film of oil which is constantly refreshed as oil is supplied to the bearing. Intuitively, the
thickness of the film is determined by the shaft load, shaft speed, and the rate of flow of
oil into the bearing as well as the viscosity of the oil. This type of lubrication is common
to journal bearings and gears found in gas turbine engine systems.

The concept of rolling-element bearings is shown in Figure 7.3 [2]. This figure adapts
the motion shown in Figure 7.1 from sliding motion to rolling motion, which in an
idealized sense is completely free of friction.

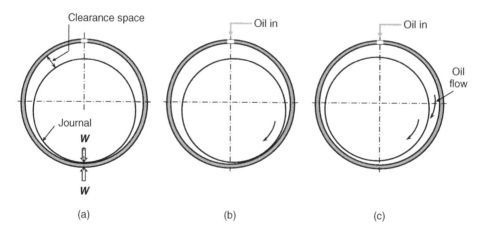

Figure 7.2   Classical journal bearing with hydrodynamic lubrication.

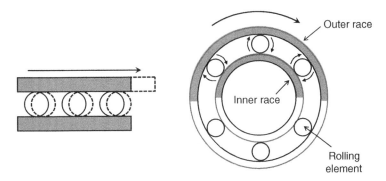

Figure 7.3 The rolling element.

In these circumstances, the load from the shaft is concentrated at the points of contact between the rolling elements and their respective races. This suggests that, at the points of contact, stresses are extremely high and localized deflection of the material of both ball and race is to be expected. In fact, the surface deflections are such that the contact becomes an area as depicted in Figure 7.4.

While not large, the deflections are sufficient to generate substantial heat and contribute to surface fatigue of the components. In the absence of any manufacturing defects, and perfectly clean lubrication, a rolling-element bearing will eventually fail in fatigue regardless of the quality of lubrication.

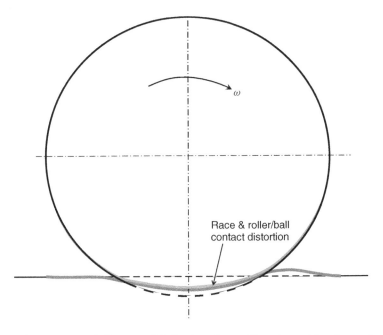

Figure 7.4 Exaggerated diagram of deflections in a rolling-element bearing under load.

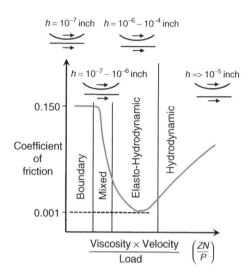

Figure 7.5 Friction coefficient as a function of velocity-viscosity/load parameter.

A complete description and analysis of rolling-element bearings is provided by Harris [3]. This document makes clear that the practical implementation of rolling-element bearings involves a number of surface motions that, in the absence of lubrication, will generate considerable frictional heating. Lubrication of highly loaded bearings is therefore essential to reduce friction and to dissipate heat.

Zoretsky [4] provides a detailed description of the lubrication regimes that span the range of possible conditions. This is conveniently described in the well-known 'Stribeck plot' shown in Figure 7.5.

Beginning in the boundary region of Figure 7.5, this might be described as the starting condition for a bearing where oil may be supplied but velocity is low and the load is relatively high. Under these conditions, even with superior finish of the bearing surfaces, there will be metal-to-metal contact and the coefficient of friction will be determined by the kinetics of interaction between asperities found in the two contacting surfaces.

In the mixed region speeds are increasing, more lubricant is made to flow at the bearing interfaces, and physical contact is rapidly being diminished. In this region, the bearing is beginning to function as designed and the coefficient of friction is falling rapidly, resulting in a much lower level of heat generation.

In the elasto-hydrodynamic (EHD) region of operation, the combination of velocity and load together with a lubricant of optimum viscosity results in the bearing fully operating on a liquid film of oil. In this region of operation, the combination of viscosity and elasticity of the lubricant allows the nominal region of bearing surface contact to be in dynamic equilibrium and friction forces are at a minimum.

Beyond the EHD region of operation, the space between the bearing surfaces increases and the amount of lubricant required to support the process is equally increased. This is referred to as the hydrodynamic region of operation and is typical of journal bearing operation. In the hydrodynamic region, the combination of load, speed, and lubricant viscosity are such that the viscous drag forces are beginning to dominate. As a consequence, the apparent coefficient of friction is increased.

Since all of the energy absorbed by frictional forces is dissipated as heat, the obvious region of operation most desirable is the EHD region. This operating region will produce the least losses and, if managed properly, will result in the longest life of the affected parts. The choice of lubricant is therefore important and the mechanisms of heat rejection must be sufficiently robust that the operation remains within the temperature limits of the lubricant chosen.

The determination of the ratio of heat generation and dissipation are fundamental to the selection of appropriate lubricants and to the design decisions on flow rates to bearings. Harris [3] provides an analysis of the heat generation process for typical bearing arrangements. This is a complex analysis task, which involves calculations of the forces and movements on each of the elements of the bearing. These, in turn, can be used to calculate the frictional power loss within the bearing. Since all of the frictional power loss ends up as heat, this heat must be removed or the temperatures within the bearing will rise to unacceptable levels. Conduction of heat away from the bearing using the surrounding metal as a conductor is a common technique in industrial machinery where bearing speeds and loads are relatively low. With the possible exception of the fan/compressor bearings in a modern gas turbine engine, all other bearings are surrounded by thermodynamic processes that are at temperatures very much higher than can be sustained in a highly loaded, high-speed bearing. The only practical method of cooling aircraft engine bearings is therefore by convection, using the lubricant as a coolant.

Let us therefore consider the problem of simultaneously lubricating and cooling a rolling-element bearing. For lubrication purposes, it is necessary to get the oil to those locations on the bearing where both rolling and sliding contact occurs.

Figure 7.6 shows the multitude of points on a typical ball bearing which require continuous lubrication. The points at which the rolling elements contact the race are the major

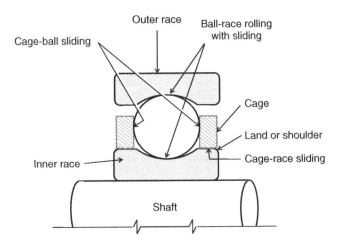

Figure 7.6   Required lubrication points in a bearing.

load-carrying points through which forces generated by the rotor are transmitted to the frame. Since they are in rolling contact, the stresses induced are primarily surface compression leading to heating and cyclic fatigue. The cage constrains the rolling elements to a fixed circumferential separation. These points of contact are therefore primarily sliding friction, which causes wear and generates more heat.

In practical designs, the oil is delivered to the various points in the bearing using pressurized jets aided by the centrifugal (slinging) effect of bearing rotation. A schematic of bearing lubrication is shown in Figure 7.7.

In this figure, the jet is delivered to a lip on the inner race of the bearing and is allowed to flow through orifices drilled in the inner race. By selecting the size of the orifices carefully, a prescribed amount of oil is delivered to the ball which is then distributed to the cage and to the outer race. As indicated in the figure, the main sliding interface between the cage and the inner race is supplied with oil by separate orifices.

Finally, the figure indicates the use of a separate annulus on the outer race, which serves two functions. First, and most fundamental, this flow of oil serves to cool the outer race. In addition, if the annulus is arranged properly it serves as a static pressurized bearing between the outer race and the engine structure. Such an arrangement provides considerable damping in the event of rotor imbalance caused by such events as a blade loss. This latter function is sometimes referred to as 'squeeze film damping'.

The detailed design of each bearing arrangement will obviously result in different combinations of oil delivery and cooling passages. However, in principal, these concepts embody the mechanisms for lubricating bearings in the gas turbine engine.

The final step in the process of lubrication and cooling is the continuous removal of oil from the bearing cavity. In industrial machinery, this can be achieved by gravity whereby

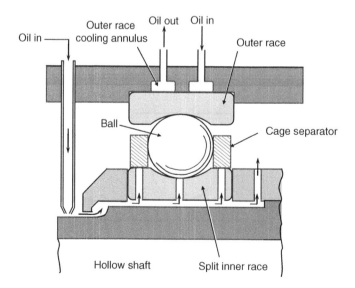

Figure 7.7 Under- and outer race cooling for bearing thermal management.

the oil pump is positioned at or within the base of the machine. In airborne systems, this arrangement is not possible for several reasons. First, the engine can assume a great range of attitudes in flight ranging from steady, level flight through to violent flight maneuvers such as rolls and dives common to fighter airplane exercises. Such maneuvers occur at many times the acceleration of gravity, making any notion of gravity drain of oil non-sensical. Secondly, there is a strong motivation to keep the lubrication processes within the EHD film regime (see Figure 7.5). This precludes flooding the bearing cavity with oil. Finally, to maximize the cooling effect, it is desirable to remove the oil as quickly and as completely as possible.

Removal of oil from the bearing cavity requires that it be pumped away for all of the reasons cited above. To ensure that this is achieved, the bearing housing is vented or pressurized with bleed air from the compressor and a scavenge pump is employed to suck away the oil and whatever air is required to ensure continuous and rapid drainage of the bearing. These are high-volume pumps which typically deal with a highly aerated mixture of air and oil. Ratios as high as 90% air and 10% oil are not uncommon.

To complete the picture of lubrication and cooling of a bearing, we need to consider the nature of the lubricant itself. In aircraft gas turbine applications it must be possible to start the engine for any ambient condition from -40 to $+50\,°C$. It is therefore customary to specify lubricants at the cold conditions in terms of a viscosity and/or a pour point. For example, a common viscosity requirement is $15\,000$ cSt (centistokes) at $-40\,°C$ and a pour point at $-54\,°C$. These numbers reflect a desire to start the engine at $-40\,°C$ without the need to preheat the oil. The specification of a pour point (the lowest temperature at which the material will still behave as a liquid) provides some temperature margin on pump ability, thus ensuring that oil can be delivered to the bearings at $-40\,°C$.

At the high end of the temperature scale, there are several attributes of the lubricant that are important to the successful lubrication and cooling of the bearings. First, the viscosity of the oil must be such that the EHD film is maintained.

Figure 7.8 provides data on a variety of lubricating oils. Mineral oils have a useful upper temperature of the order $200\,°C$ whereas several of the synthetic oils indicate a temperature range from -40 to $+300\,°C$. While mineral oils can operate at the higher temperatures, their stability at these temperatures dictates more frequent oil changes.

A major attribute of lubricants at elevated temperatures is their oxidative stability. Coking is the more severe form of oxidation, forming soot which is undesirable for many reasons (not least of which is its ability to block oil galleries). At elevated temperatures however, oils will begin to break down and form free radicals. This changes their viscosity, making them less effective as a lubricant.

The final attribute of a lubricant to be considered here is its capacity to avoid creating foam. Aeration during the scavenge process is inevitable, but non-foaming oils will separate very much more quickly from entrained air and revert to liquid form than a lubricant with a strong tendency to foam.

The ester-based oils are the commonly accepted lubricant for gas turbine engines. Zoretsky [4] suggests that the characteristics defined in Table 7.1 establish the requirements for an advanced turbine engine high-temperature lubricant. He does not state that available oils fully meet this requirement, but lubricant developers consider Table 7.1 to be a worthy goal.

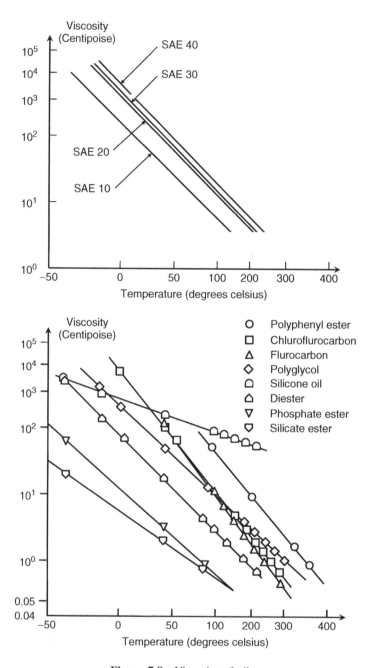

Figure 7.8   Viscosity of oils.

Table 7.1 Requirements for an advanced turbine engine high-temperature lubricant.

Physical property	Condition or limit
Physical property	<15 000 cSt −54 °C (−65 °F) >1.0 cSt at 260 °C (500 °F)
Compatibility with other materials	–
Oxidative stability (potential bulk oil temperature)	260–427 °C (500–800 °F)
Evaporative loss after 6.5 hours at 260°C (500 °F)	<10%
Lubricating ability	Satisfactory after 100 hours bearing rig test at 260–316 °C (500–600 °F) tank temperature (USAF) specification
Flashpoint (minimum)	260 °C (500 °F)
Pour point	−54 °C (−65 °F)
Decomposition	No solid products or excessive deposits in 100 hour bearing test
Foaming	Non-foaming

7.2 Lubrication System Operation

Any lubrication system is a closed-flow system driven by requirements for oil flows at specific pressures. These flows are in fact determined by bearing lubrication needs which, in turn, are dictated by EHD film thicknesses and cooling requirements.

For example, EHD film thicknesses for typical bearing loads and speeds require lubricant viscosities of the order 5 cSt. Figure 7.8 offers a number of lubricants in this range; however, the ester oils are commonly used because of their superior temperature range. The figure suggests that the operating temperature of the oil should be of the order 150–175 °C in order to provide the necessary EHD film thickness.

Considering the cooling requirements, prudent design practices suggest that the oil be delivered to the bearing via jets located in the bearing housing and that this oil is scavenged away as quickly as possible. The scavenge pumps are sized such that they can handle an air/oil mixture of the order 9:1. This further suggests that the residency time of the oil in the bearing housing is relatively short. Again, referring to Figure 7.8, it would seem prudent to limit the rise in oil temperature to about 25 °C. While this will ensure that its rate of oxidation is kept under control, it will also contribute to keeping oil system heat exchangers to a modest size.

The quantity of frictional heat generated by each bearing is a function of its size, speed, and loading. Harris [3] describes the calculation methods to be employed in obtaining the frictional loads on a rolling-element bearing. Using a #218 angular contact bearing as an example, he calculates the heat generated for a range of loads at a fixed speed of 10 000 rpm. These data are presented in Figure 7.9. Using 15 000 BTU/h as the operating point and limiting the oil temperature rise to 25 °C yields a required oil flow of 12 lb/min.

The foregoing analysis must be conducted for all of the bearings in the system. The total oil flow thus calculated defines the size of supply pump that will be required. A single supply pump is usually employed to deliver oil to each bearing housing and to the

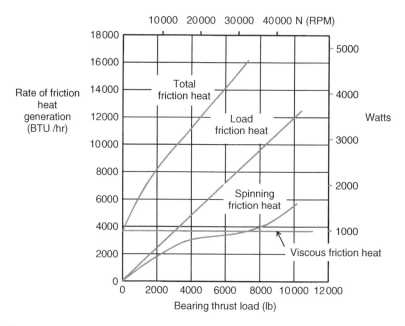

Figure 7.9   Frictional heat generation for a #218 angular contact ball bearing at 10 000 rpm, 5 cSt oil viscosity with jet lubrication.

accessory gearbox. By contrast, a separate scavenge pump is usually provided for each bearing housing. This arrangement is partly due to the high volume of air (typically 9:1 air/oil ratio) that must be handled and partly due to the greater difficulty of balancing a system for which the size of the vent, together with flight conditions, dictates the quantity of air entrained with the oil.

7.2.1   System Design Concept

A schematic diagram of a typical lubrication system is shown in Figure 7.10. The principal features of the typical lubrication system are described in Sections 7.2.1.1–7.2.1.6.

7.2.1.1   Oil Tanks

The oil tank provides a reservoir of oil available to the system. This provides the necessary spare capacity to accommodate some oil consumption and the physical space necessary to house the de-aerator.

There are no set rules for the location of the oil tank and it may or may not be fitted to the engine. In large modern turbofans, it is very much part of the engine and will almost certainly be mounted on the external cowling of the fan portion of the engine where the environment is the most benign.

By contrast, on a small turboprop engine the oil tank will likely be mounted on the airframe. While specified by the engine manufacturer, it may be supplied as part of the aircraft.

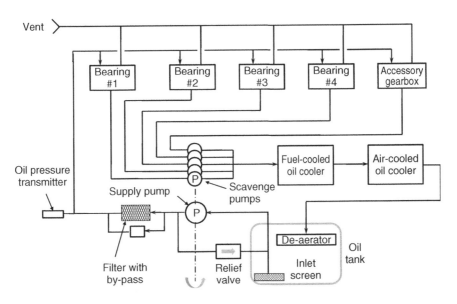

Figure 7.10   Typical lubrication system schematic.

7.2.1.2   Tank Pressurization

In most commercial applications of the jet engine, the oil tank is simply vented to atmosphere. In some military applications, however, it is common to close the tank to the atmosphere at altitudes above 30 000 ft. This is accomplished using a simple bellows/spring poppet valve as shown in Figure 7.11.

The bellows spring is pressurized to a preset value, which is less than the atmospheric condition at sea level. At ground level and up to a certain altitude, the tank is therefore vented. At an altitude where atmospheric pressure equals the pressure inside the bellows, the valve begins to close. As further altitude is gained, the valve closes fully and the tank is sealed at the atmospheric pressure associated with that altitude. This is sufficient to ensure that the boost pump does not cavitate at the prevailing suction conditions. It may also be necessary to fit the tank with a pressure relief valve to prevent overpressurization of the oil tank. This is particularly important where the bearing cavities are pressurized by compressor bleed air.

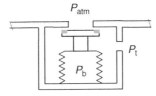

Figure 7.11   Typical lube tank vent system.

7.2.1.3 Oil De-aeration

The bearing cavities are either vented to the atmosphere or pressurized with bleed air extracted from the engine compressor. In either case, as the scavenge pumps remove the oil from the bearing housings, large quantities of air are entrained which must be removed from the oil. The oil tank is therefore fitted with a de-aerator which strips the oil of entrained air as it is returned to the tank. This is achieved by centrifuging the returned mixture to bring the heavier oil out to the circumference of the centrifuge, thereby providing a central path for the air to escape. The centrifuged oil is then delivered to one of several horizontal trays where nearly all of the momentum of the flow is dissipated and the oil, under gravity, migrates to the bottom of the tray. The air is free to escape from the oil surface and through the vent to atmosphere. It is clear from this argument that a closed tank used in military applications will require a pressure relief valve to avoid possible overpressure conditions.

7.2.1.4 Oil Supply and Scavenge Pumps

Oil is delivered to the engine bearings via a high-pressure pump which receives de-aerated fuel from the oil tank. The oil supply pump acts as the primary oil supply to all of the engine bearing sumps and to the accessory gearbox.

These pumps are usually positive displacement types with gear pumps, gerotor pumps, and vane pumps in common usage today. In all cases, the lubrication pumps are driven mechanically via the accessory gearbox. The pumps therefore operate at some multiple of engine speed and, if a relief valve is not fitted, provide flow and pressure which varies with engine speed. It is therefore common practice for the pump to be fitted with a pressure relief valve, thus ensuring that an overpressure condition cannot occur. Excess oil from the pressure relief valve is returned to the oil tank.

As indicated in the system schematic of Figure 7.10, there are several scavenge pumps which remove oil from each bearing sump as well as the accessory gearbox sump. This oil is then discharged back to the tank via the oil de-aerator system. Typically, the oil supply and scavenge pumps are installed in a common unit with a single spline-drive connecting all of the pumping elements to the accessory gearbox.

Figure 7.12 is a photograph of the lubrication and scavenge pump unit used on the Allison (now Rolls-Royce) AE3007 turbofan engine. In this particular application, gerotor pumping elements are used for both the high-pressure supply and bearing cavity scavenge functions.

7.2.1.5 Oil Filter

An oil filter is provided either as an integral component of the oil tank or as a separate component; the oil is therefore filtered before delivery to bearing system. The choice of filtration level has received considerable attention in recent years. In older systems, it was common practice to utilize barrier filters with the capacity to trap debris greater than $25\,\mu m$. As more became known about the reduction in bearing life as a consequence of over-rolling hard debris (a $3.5\,\mu m$ particle of silica can cause a minute crater in a heavily loaded bearing which, in turn, can initiate a point of surface fatigue), the industry began to adopt the practice of finer filtration. Today, $10\,\mu m$ filtration is common and there have

Figure 7.12 Rolls-Royce/Allison AE3007 lubrication and scavenge pump (courtesy of Parker Aerospace).

been trials using $3\,\mu m$ filtration. As may be expected, the finer filtration demands higher delivery pressures so that pump designs are directly affected.

From Figure 7.10 note that a pressure differential sensor is fitted to the filter housing, which indicates increases in filter blockage. An integral component of this type of system is a filter bypass circuit, which begins to allow oil to bypass the filter in order to ensure oil is delivered to the bearings in the event of partial to complete blockage of the filter. This is the first of several system protective devices that are fundamental to overall protection of the engine.

7.2.1.6 Pressure Monitoring

In addition to signals from the pressure differential measurement at the filter, overall oil pressure is sensed and both are transmitted to electronic monitors and/or cockpit displays. In land/marine applications of the gas turbine, a low oil pressure indication would trigger a system alarm and shut down. An airborne system flight safety precludes such an action and only the alarm is used.

The filtered and pressurized oil is routed to each of the bearing housings. As noted previously, the oil is typically delivered to the bearing through a series of jets. Such a design requires to be balanced to ensure that adequate oil flow is delivered to each bearing housing. As noted in Figure 7.10, a restrictor may or may not be required in order to ensure balanced delivery to each bearing system. It should also be noted that the selection of the jet sizes within a bearing housing further contributes to the even distribution of oil to all required points of lubrication.

7.2.2   System Design Considerations

A casual review of the major elements that comprise the oil delivery system indicate that it is, from the designer's perspective, a piping network. Piping network calculation schemes are useful in initial sizing of pumps, piping, and other components; however, the relentless requirement for minimum weight dictates that the designs is also supported by rig tests which contribute to libraries of system-specific component performance. They are also the final arbiters in ensuring that the design is adequate. Finally, they are also the means by which the system is balanced and tuned.

7.2.3   System Monitoring

The inclusion of a variety of measurement in a lubrication system reflects the criticality of this system to flight safety. In general, these measurements fall into several key categories as described in Sections 7.2.3.1–7.2.3.7.

7.2.3.1   Oil Pressure

System monitoring is performed to ensure that the lubrication system is performing as intended. There are a variety of useful measurements for this purpose.

   In order to ensure that oil is flowing to the bearing systems, it is normal practice to fit the lubrication system with a pressure transducer downstream of the main supply pump. This transducer will be monitored continuously, typically by the engine full-authority digital electronic control (FADEC) controller, which provides engine health status to the flight crew via a display system. In Boeing aircraft this display system is called the 'engine indication and crew alerting system' (EICAS). In Airbus Aircraft the equivalent display system is called the electronic centralized aircraft monitor (ECAM). In the event of a loss of pressure, the flight crew will receive a warning message requiring that the engine be shut down.

7.2.3.2   System Temperature

All oils have viscosity variation with temperature and viscosity is the principal parameter controlling the film thickness within the bearing. We therefore need to be sure that the oil is maintained within appropriate limits. As with the pressure measurements, a temperature sensor may be fitted to the oil tank or to the oil line downstream of the supply pump. The choice of location is dictated by the arrangement of the major components of the system. Figure 7.10 indicates that the oil coolers (oil/oil and fuel/oil) were placed after the scavenge pump and before returning to the tank. This arrangement must therefore deal with the large amount of entrained air within the coolers, suggesting that they will be larger and heavier. In such an arrangement, the temperature sensor will likely be placed in the tank.

It is also possible to design for operation with a 'hot' tank. In hot tank designs, the coolers are arranged downstream of the pump. The de-aerator, which is a component of the tank, strips away the air making it necessary to select pumps and filters capable of handling the hot oils. The coolers, which are placed downstream of the pump and likely the filter, are now designed for liquid only and can be more compact and thermally efficient. The most appropriate location for a temperature sensor would therefore be downstream of the oil cooler system.

7.2.3.3 Condition Monitoring

The purpose of condition monitoring is to provide information considered to be vital to the functional status of the system, for which maintenance action must be taken either immediately or within some predetermined time window.

7.2.3.4 Filter Blockage

There are two elements to the problem of the oil filter blockage. The first is the measurement of pressure drop across the filter, which may or may not be provided as information to the flight crew. The second is the provision of a filter bypass valve, which operates once the pressure drop exceeds some safe limit.

As a rule of thumb, the pressure differential measurement will indicate alarm conditions when its value falls below 15% of the incoming supply pressure. Typical operating pressure is of the order 200 psi, suggesting pressure drops of the order 30 psi should not be exceeded.

Similarly, at a pressure drop of the order 45% of supply pressure the filter is, for all practical purposes, blocked. This condition will cause the bypass valve to open, thus bypassing the blocked filter and providing vital lubricant to the bearings. Although inadequately filtered, system safety dictates that it is better to continue to supply oil to the bearing than to risk immediate failure from oil starvation.

7.2.3.5 Oil Debris Monitoring

The bearing system of a gas turbine engine is absolutely vital to the safe operation of the engine. While the engine is designed to survive a bearing failure in flight, it is a condition which demands an immediate shutdown of the engine.

Following the logic that if the system operates with an EHD film 100% of the time and the oil is maintained at a high level of cleanliness, a bearing could only fail in surface fatigue. In fact, statistical analysis suggests that the occurrence of bearing failure follows the causes listed in Table 7.2.

Mechanical damage is usually associated with minute flaws or damage caused by installation of the bearing. The imposed surface imperfection results in early surface damage.

Corrosion is associated with some form of chemical etching of the surface of either the race or the rolling element. Again, it initiates early surface damage.

The presence of hard debris in the oil will most likely result in it being trapped between the race and the rolling elements of the bearing. Figure 7.5 indicates that the film thicknesses in the EHD range of lubrication are typically $10^{-6}-10^{-4}$ inches and that a 3 μm

Table 7.2   Principal causes of bearing failures in aircraft engines.

Cause of bearing failure	Frequency of occurrence (%)
Mechanical design	30
Corrosion	30
Over-rolling debris	30
Metal fatigue	10

$(1.2 \times 10^{-4}$ inch) particle falls into this range. Since many of the older systems are filtered to only 25 µm, it follows that bearing damage is likely to occur if the debris is sufficiently hard to dent the surface of the bearing race and/or rolling element.

We can therefore conclude that, regardless of the cause of the failure, failure of a bearing is always associated with damage resulting in metallic particulate being generated, with considerable likelihood that it will be scavenged away from the bearing.

Figure 7.13 shows the history of a bearing failed intentionally during rig tests. The figure indicates the quantity of debris possible from a large turbofan thrust bearing and the likelihood that the failure will progress to self-destruction in a relatively short period of time. At DN numbers of the order 2×10^6 it is expected that the bearing will initiate an engine shutdown in approximately 150–200 hours after the initial spalling damage occurs. (A commonly used measure of bearing performance is the so-called DN number which is defined as the bore diameter in centimetres D multiplied by the rotational speed N in revolutions per minute.)

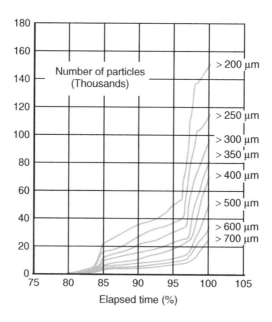

Figure 7.13   Failure progression of a thrust bearing versus time.

7.2.3.6 Chip Detectors

The use of magnetic devices as a means of capturing debris in an oil line has been established for more than 50 years. A typical magnetic plug (or 'mag-plug') is shown in Figure 7.14. This is a passive device, which simply captures and holds the debris that it has successfully removed from the flow. Once captured, the debris must be inspected and analyzed and a decision made about its meaning.

In general, the magnetic plug is a low-cost device that can provide considerable insight into the state of the bearings if it is analyzed by a competent individual. In a straight-line installation as shown in Figure 7.14, the magnetic plug has a capture efficiency as low as 5%. By installing the device at a right-angle bend, the capture efficiency can be greatly improved.

The principal criticism of the magnetic plug is that is has no convenient and reliable indication of the presence of metallic debris. A technician must physically remove the plug and conduct an analysis. In its early use in situations involving trained technicians, the device was a reasonably successful tool. In more recent years, where aircraft are deployed around the world, the use of the same technician is not possible and the variability of interpretation from individual to individual has proved problematic.

The incorporation of two electrical contacts into the plug such that the arrival of metallic debris could form an electrical bridge between the contacts has been in use for years. The closure of the contacts allows the triggering of an alarm. Such an alarm can be provided to the pilot, allowing them to take appropriate action.

In general, the combination of a magnetic plug with electrical contacts has introduced as many problems as it claims to have solved. A small amount of very fine debris can close the contacts. This material may be obtained from normal wear and may have been circulated to the mag-plug from other parts of the engine leading to false positive indications of a problem. This state of affairs has led to the introduction of so-called 'fuzz burners'. These devices send a high current through the contacts, which tends to burn off fine debris in much the same way that high current will burn out an electrical fuse. Using the concept, it is argued that if the debris is significant it cannot be burned away and further inspection is required.

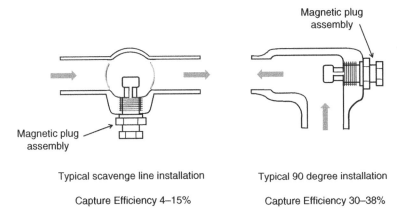

Typical scavenge line installation   Typical 90 degree installation

Capture Efficiency 4–15%   Capture Efficiency 30–38%

Figure 7.14 Magnetic plug installation examples.

There have been a number of spectacular accidents involving loss of life due to pilots choosing to ignore an electrical chip detector. The reader must choose for himself the wisdom of incorporating an electrical chip detector as part of a design.

7.2.3.7 Inductive Debris Monitors

The use of inductive coils to generate a magnetic field around or in the oil line has proven to be a more sophisticated and reliable means of providing an indication of bearing distress. A schematic of such a device is shown in Figure 7.15.

The basic operation of the device involves generating opposing magnetic fields in the outside coils and placing a third sense coil at the point of neutrality. A conductive particle will, in traversing the pipe, disturb the balance of the two fields which is detected by the sense coil.

Such a device is capable of detecting the passage of debris as small as 100 μm. It is now known that a substantial number of particles larger than 100 μm are generated even from a small bearing (e.g., Figure 7.13). It is therefore desirable to accumulate a number of counts before sounding an alarm.

The fitment of an oil debris monitor poses some questions of complexity. Industrial gas turbines which, for the most part, operate unmanned have opted to fit an oil debris monitor to each scavenge line. Such an arrangement allows remote monitoring and a clear indication of the origin of bearing distress.

For airborne installation, weight and system complexity are important factors. It is therefore common practice to place the device at a point in the scavenge system where all of the individual scavenge flows have been combined. This is typically downstream

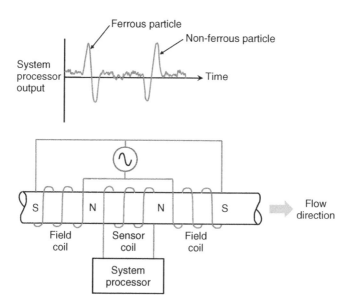

Figure 7.15   Oil debris monitor (courtesy of GasTOPS Ltd).

of the scavenge pump system. Individual magnetic plugs can then be fitted to individual scavenge lines which allow diagnosis down to individual bearing cavities.

A modern oil debris monitor has associated electronics which allow communication with engine control systems and/or an electronic diagnostic unit either specific to the engine or part of a larger aircraft system.

7.2.4 Ceramic Bearings

At the end of this chapter, it is relevant to address an interesting new technology that is the result of a long-term effort within the industry in pursuit of bearing materials whose hardness exceeds that of heat-treated M51. The criticism of all rolling-element bearings is the deformation of the rolling element under load, leading to greater heat generation. Modern lubricating oils have a heat removal function which is equally important to its lubrication function.

Work has progressed on ceramic bearings to the point where many demonstrations have occurred; however, failures modes are rather abrupt. When failures do occur, the associated debris has the same strength and sharpness as a cutting tool. It has therefore been said repeatedly that the major impediment to its use is the inability to detect failures early enough for safe operations. Nevertheless, work continues on the development of alternative bearing materials and it is very likely that such bearings will see service in the future. It is reasonable to predict that such bearings will first see service in less safety-critical applications than in an aircraft propulsion plant; however, it seems inevitable that materials research will eventually yield an aeronautical class bearing.

References

1. Faires, V.M. (1965) *Design of Machine Elements*, 4th edn, MacMillan Company, New York.
2. Anon. (1986) *Rolling Bearings and their Contribution to the Progress of Technology*, FAG, Schweinfurt.
3. Harris, T.A. (1991) *Rolling Bearing Analysis*, John Wiley & Sons, Inc., New York.
4. Zaretsky, E.V (ed.) (1997) *Tribology for Aerospace Applications*, Society of Tribology and Lubrication Engineers, STLE-SP-37.

8

Power Extraction and Starting Systems

The typical aircraft propulsion gas turbine is used as a source of mechanical power for many of the systems associated with both the engine itself and the airframe. Mechanical power must also be applied to the engine's rotating machinery during the start process.

In addition to mechanical power transfer, bleed air from the high pressure (HP) compressor is used as a source of power for a number of airframe and engine systems.

The following sections describe the various power transfer systems and equipment used on both commercial and military aircraft together with any systems implications involved.

8.1 Mechanical Power Extraction

Transfer of mechanical power from the engine (for engine and airframe systems usage) and to the engine (for starting) is accomplished via the accessory gearbox which is connected to the HP shaft of the engine. This is accomplished using a tower shaft connected to the HP shaft via a bevel gear and, at the opposite end, to an accessory gearbox mounted either on the core of the engine or on the fan case for some turbofan applications (as mentioned previously in Chapter 3).

Military aircraft often utilize a separate airframe-mounted accessory gearbox as illustrated in Figure 8.1 which shows a top-level overview of the arrangement employed by many US and European fighter aircraft including F-15, F-18, B-1B, and Eurofighter (Typhoon). In this approach, the accessory drive is considered to be part of the airframe.

Whether an airframe-mounted accessory drive (AMAD) or a more traditional engine-mounted accessory gearbox arrangement is employed, the related systems equipment that is mounted on the accessory gearbox is similar and can be represented by the schematic of Figure 8.2; see also [1].

8.1.1 Fuel Control Systems Equipment

Referring to Figure 8.2, the engine fuel pump power extraction includes the gas generator HP pump and, where an afterburner is employed, an additional pump is provided. The

Gas Turbine Propulsion Systems, First Edition. Bernie MacIsaac and Roy Langton.
© 2011 John Wiley & Sons, Ltd. Published 2011 by John Wiley & Sons, Ltd.

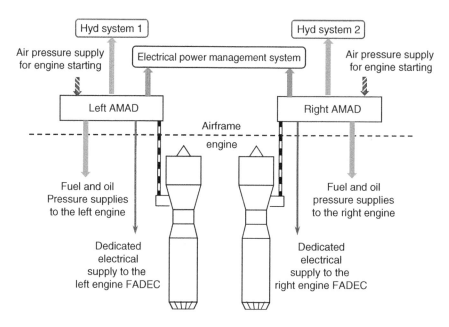

Figure 8.1 Overview of an airframe-mounted accessory drive arrangement.

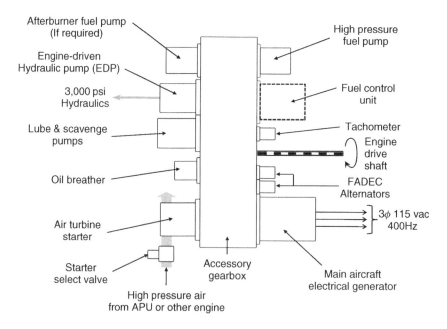

Figure 8.2 Typical accessory drive power transfer sources.

afterburner pump operates at much lower pressures than the gas generator HP pump and is typically a centrifugal type of design. The afterburner pump may be driven mechanically via the accessory gearbox as shown or, alternatively, via a bleed-air-powered turbine arrangement (see Section 8.3).

The fuel control unit, shown mounted to the accessory gearbox in the figure, is typical of the traditional hydromechanical devices which include acceleration and deceleration limiting, speed governing, and metering functions; an engine speed input is therefore required. The amount of power absorbed by this device is negligible. The dotted line shown indicates that, for the modern full-authority digital electronic control (FADEC) engine, a simpler fuel metering unit (FMU) replaces the fuel control. This device may be mounted off the gearbox since all the control functions are provided by the FADEC.

An engine speed sensor, either a tachometer or magnetic pulse pick-up, is required to provide independent speed information for crew. This device is usually installed on the gearbox.

For FADEC-controlled engines, a redundant power supply is provided via two permanent magnet alternators (PMAs), driven by separate drive pads, in order to support the functional integrity requirements of the fuel control system. Shaft speed information can also be derived from the PMA waveform for control purposes.

In some smaller aircraft applications such as regional and business aircraft, an additional fuel pump may be required to provide fuel pressure to the airframe to power fuel feed and/or scavenge ejector pumps associated with the airframe fuel system [2]. This pump is typically limited to 300–500 psi above airframe fuel boost pressure and the output may include a flow fuse to shut off the pump output in the event of a break or major leak in the line between the engine and the airframe which could otherwise lead to an engine fire.

8.1.2 Hydraulic Power Extraction

The primary hydraulic power source for airframe flight controls actuation and utilities such as landing gear extend/retract wheel brakes and steering is provided by engine-driven hydraulic pumps (EDPs). This is accomplished via variable displacement piston pumps that maintain a nominal system pressure in the range 3000–5000 psi. Commercial aircraft today employ (almost exclusively) a 3000 psi standard for aircraft hydraulic equipment. Some military aircraft have 4000 psi hydraulic systems in order to reduce hydraulic system (and hence aircraft) weight. In a few cases, system pressures of over 5000 psi have been adopted. However, the benefits of higher system pressures offer continually diminishing returns due to the fact that flight control actuator stiffness (a key factor in flutter margins) is directly proportional to actuator area. When considering flight control surface actuation, the benefits of pressure times area as an available force cannot be fully realized as a means of achieving system weight reduction.

These pumps utilize a swash-plate type of arrangement to vary the displacement as indicated by the schematic of Figure 8.3. The swash-plate actuating servo modulates the swash-plate angle as a function of delivery pressure. The pump may also incorporate a device that sets the swash-plate in the neutral position in order to minimize pump torque during the engine-start phase of operation.

In some applications, two EDPs may be installed on the accessory gearbox in order to meet the functional integrity of the hydraulic system and its associated loads.

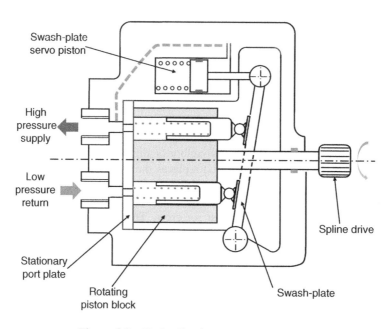

Swash-plate
servo piston

High
pressure
supply

Low
pressure
return

Spline drive

Stationary
port plate

Rotating
piston block

Swash-plate

Figure 8.3   Hydraulic piston pump schematic.

8.1.3   Lubrication and Scavenge Pumps

The engine lubrication and scavenge system comprises a HP oil supply pump and as many as five scavenge pumps, typically installed within a single housing and sharing the same mechanical drive. Fixed displacement pumps are employed for this function and gear, gerotor, and vane-type pumps are in common use today. A maximum oil delivery pressure of about 500 psig is typical for lubrication system oil supply pumps.

8.1.4   Electrical Power Generation

One of the most critical power extraction requirements for the aircraft propulsion engine is to provide electrical power to the aircraft. For most engine applications, electrical power is generated via an engine accessory gearbox-mounted generator which provides a 400 Hz alternating current three-phase supply with a nominal voltage between phase and neutral of 115 V.

The challenge for the electrical power generation system is to accommodate the engine speed variation (and hence the generator drive speed) between idle and maximum, which is typically about 2:1 while delivering alternating current (AC) power at a constant frequency in accordance with the power generation specifications.

There are three approaches that address this fundamental problem.

1. Introduce a constant speed drive (CSD) between the engine drive shaft and the generator.

2. Shape the output waveform electrically to provide a variable speed constant frequency (VSCF) generator output.
3. Generate variable frequency power and have the electrical power users on the aircraft manage the resulting 2:1 frequency variation.

The CSD approach employs a hydromechanical device that is driven by the variable speed accessory gearbox and outputs a constant speed suitable for driving the electrical generator. Figure 8.4 shows a top-level schematic of such a device.

As shown, the CSD contains a differential transmission gear with a rotational speed output equal to the sum of two inputs. One of these inputs is the accessory gearbox drive; the second input to the differential is via a hydraulic piston pump/motor arrangement. One of the hydraulic units is a fixed displacement device while the other is a variable displacement unit whose swash-plate angle is varied by a servo actuator. The position of the servo actuator and hence the swash-plate angle is determined by a flyweight type of speed governor.

When the accessory drive speed is below the required output speed, the hydraulic drive system adds its speed to the differential input; in this mode the variable displacement unit acts as a pump.

Conversely, when the accessory drive is higher than the required output speed, the hydraulic unit rotates in the opposite direction thus subtracting its speed from the accessory drive speed. In this mode of operation, the variable displacement unit acts as a motor.

The hydraulic section of the CSD is quite complex with additional items not shown in the schematic for clarity. These items include a filter with pressure-drop indicator, a reservoir, and a scavenge pump. The hydraulic oil is cooled typically via a separately located oil cooler.

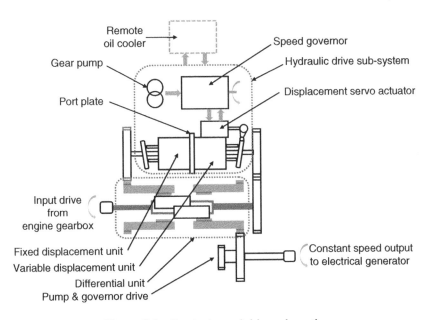

Figure 8.4   Constant speed drive schematic.

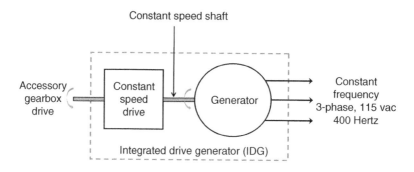

Figure 8.5 Integrated drive generator concept.

The typical CSD also contains a clutch that can be selected by the flight crew following a drive malfunction to disconnect the drive from the accessory gearbox. This clutch can only be reset on the ground by maintenance staff.

A relatively recent development of the CSD/generator arrangement is the integrated drive generator (IDG) which combines the CSD and generator into a single line-replaceable unit (LRU) as shown in the conceptual drawing of Figure 8.5.

The VSCF approach to constant frequency AC power generation involves two different methods as described below.

- **The DC link:** In this method, raw power is converted to an intermediate DC power stage before being converted via solid-state power switching and filtering into three-phase AC power with a constant frequency of 400 Hz.
- **The cycloconvertor:** This technique uses a completely different principle. Six phases are generated at a relatively high frequency (>3000 Hz). Solid-state switching between these multiple phases in a predetermined and controlled manner has the effect of electrically commutating the input to provide the required three phases of constant-frequency 400 Hz AC power.

The DC link approach is used in many applications, usually in small capacity applications (75 kVA or less) or as a back-up to the primary generation system. The cycloconvertor technique has been used successfully on a number of US military applications including the U-2 reconnaissance aircraft, the F-18 US Navy Hornet, and the F-117 Stealth Fighter. As yet, no commercial applications employ this power generation technique.

The main problem with the VSCF approach is that all of the power conversion electronics must be rated at the full power capacity of the machine. These devices can be large and expensive with significant failure rates.

Even though the CSD and IDG approaches to constant-frequency power generation are complex and challenging from a maintainability and reliability perspective, they remain somewhat better than their VSCF counterparts.

The third approach to AC power generation, that is, the variable-frequency approach, has gained popularity over the past 15–20 years with many new aircraft program starts employing this power generation methodology. The generating system is now greatly simplified with the elimination of the CSD and its associated reliability and maintainability problems. Users of this 'wild frequency' power supply must now accommodate the frequency variations within their equipment (typically 350–800 Hz). In many applications, this results in very low power factors (below 0.5) at one of the frequency extremes with a resultant penalty in current carrying capacity and hence weight. Nevertheless, at the aircraft level, the variable frequency approach to power generation appears to yield a positive benefit.

8.2   Engine Starting

The most common method of gas turbine engine starting used today is via the air turbine starter which uses HP air from either:

- the Auxiliary Power Unit (APU) load compressor;
- bleed-air crossfeed from another operating engine; or
- HP air from a ground cart in the event of an inoperative APU.

The starter unit accelerates the engine until the gas generator speed reaches its self-sustaining speed from where it can continue to accelerate to the ground idle speed. Light-off occurs at approximately 15% speed; however, the starter must continue to assist the engine up to and beyond the self-sustaining speed before power is removed from the starter. This may take up to a minute or more and is dependent upon the prevailing ambient conditions.

The typical air turbine starter comprises:

- an air turbine wheel that generates a high-speed low-torque output;
- a reduction gear to provide a high-torque/low-speed input to engine HP shaft via the accessory gearbox; and
- a pawl and ratchet clutch that connects the starter reduction gear output to the engine.

The starter rotating assembly disengages automatically from the engine when engine speed exceeds starter output speed.

Figure 8.6 shows a schematic of an air starter together with a section through the pawl and ratchet clutch.

During the start process, it is important to ensure that turbine gas temperature (TGT) does not exceed predetermined limits so that damage to the hot section of the engine does not occur. This issue is due to the fact that, at low engine speeds, the airflow through the machine is very low and TGT will rise very quickly during the acceleration phase. Figure 8.7 shows how speed and TGT respond during the start process. Igniters are switched on as soon as the starter is engaged and switched off after the starter is disconnected.

In many gas turbines the engine start process is controlled and monitored by the flight crew; however, state of the art engines today have fully automated starting capability where the FADEC controls and monitors the process.

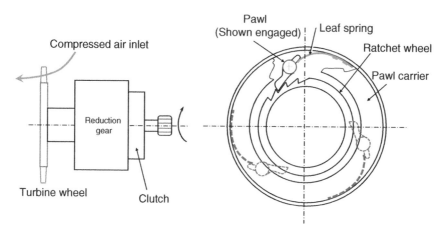

Figure 8.6   Air turbine starter concept.

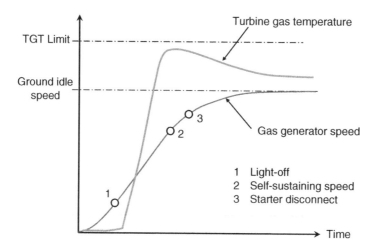

Figure 8.7   Typical speed and turbine gas temperature responses during starting.

A typical manual start sequence is as follows.

1. Select the engine master switch ON; this process switches the fuel boost pump(s) on in the aircraft fuel system.
2. Select engine starter ON to begin cranking the engine.
3. Select ignition ON.
4. As the engine speed passes 10% speed, set the power lever to IDLE. This opens the HP shut-off valve in the fuel control, allowing pressurized fuel to reach the primary fuel nozzles.

5. As the engine lights-off, signified by a rise in TGT (or exhaust gas temperature EGT), the crew must monitor temperature and be prepared to abort the start if limits are exceeded.
6. As the engine approaches idle speed, select starter and ignition OFF.

In the event of a hot start (TGT over-temperature) or a hung start (engine speed stalls before reaching idle), the HP fuel shut-off valve must be closed (power lever to OFF) and the starter should remain engaged for at least 30 seconds to purge pooled fuel and fuel vapors from the combustor.

While the air turbine starter remains the most common technique for engine starting today, many other methods have been used. In many cases, alternative starting solutions are driven by unique operational requirements; however, the following two examples are quite common and are worth mentioning here for completeness.

• Electric motor starters (and more recently the electric starter-generator) are in common use for small propulsion engines (less than 1500 HP) and for APUs. The benefit of the starter-generator is that a single unit can provide the torque input to the engine for starting and, when the engine is up and running, the same device (utilizing different windings) can be used to generate direct current (DC) electrical power for use by the engine and airframe. For starting, 28 V dc power can be obtained from either the aircraft battery of from some ground dc power supply.

   The switched reluctance starter-generator has been the subject of much research and development over recent years, and is likely to be the technology of the future for large engine starting and power generation.
• Cartridge or solid propellant starters are used almost exclusively in military applications and have the advantage of very high torque/weight ratio. The cartridge start system is also self-contained and independent of any outside power source, which may be critical in a remote hostile environment. Following a start command, the cartridge propellant is ignited. This produces high-temperature gases (typically 2000 °F) which are used to develop torque via a turbine in much the same way as the air turbine starter.

In addition to the methods described above many other techniques have been used to start gas turbine propulsions engines including hydraulic starters, liquid mono-propellant starters and, perhaps the most complex arrangement, gas turbine starters, however these solutions are very application-specific and not in common use today in either commercial or military applications.

8.3 Bleed-air-powered Systems and Equipment

In addition to the need to transfer mechanical power to and from the engine as described above, bleed air is also used as a power source for a number of engine and aircraft systems as described below.

• The afterburner fuel system in military aircraft applications may utilize a bleed-air-driven turbo-pump drive as an alternative to a mechanical afterburner pump-drive approach.

- The water injection system may also employ a turbo-pump drive where direct combustor injection is used.
- The aircraft environmental control system (ECS) together with the cabin/cockpit pressurization system requires a source of HP air.
- The aircraft anti-icing systems use hot bleed air to prevent ice build-up on wing leading edges, nacelle inlets, and engine inlet guide vanes.
- The fuel tank inerting system requires HP high-temperature air to support the air separation system which generates nitrogen-enriched air to inert the fuel tank ullage (the space above the fuel).

In some aircraft applications, bleed air has been used to power air motors driving hydraulic pumps in addition to direct mechanical or electric motor drive, in order to provide dissimilar redundancy and hence better system integrity. However, this is not a common practice in new aircraft applications. Instead, the trend is to minimize or even eliminate the use of bleed air from propulsion engines in order to improve propulsive efficiency and reduce undesirable emissions.

In addition to the primary bleed-air-powered systems summarized above, bleed air is used to provide a number of ancillary functions such as shaft bearing sump sealing and cooling. The schematic of Figure 8.8 shows a typical shaft bearing with labyrinth seals fed with bleed air.

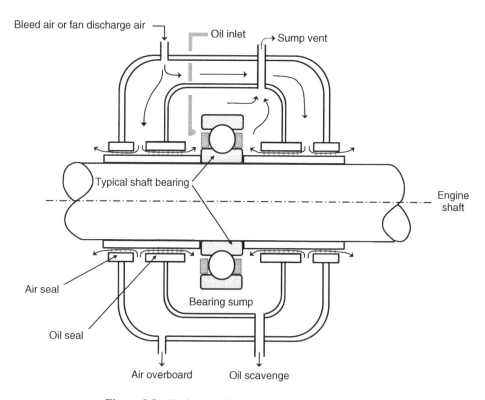

Figure 8.8   Typical shaft bearing sealing arrangement.

In many commercial high-bypass engines, fan discharge air may also be used instead of bleed air. As shown, the bleed air passes through two sets of labyrinth seals. Air flow through the outer seal goes to outside ambient. Air flow through the inner seal goes into the bearing sump which is vented to atmosphere, and the sump oil and entrained air are scavenged from the sump back to the main lubrication system tank via the de-aerator. See Chapter 7 for more detail regarding the management of bearing sump bleed air and lubrication systems in general.

In high performance gas turbines, bleed air is used for HP turbine stator and blade cooling (in order to minimize thermal stress and damage due to the extremely high TGTs that can occur) and therefore to maximize hot section operational life at these extreme operating conditions.

Stator and rotating blades are cooled by porting bleed air through hollow internal sections of the blades, thus providing convective cooling. In some designs, exits through fine holes in the blade surfaces form a cooling layer of air around the blade surface. This blade-cooling technique is called 'film cooling'.

The following section describes the primary bleed air loads typical of many commercial and military propulsion systems in service today.

8.3.1 Bleed-air-driven Pumps

Bleed-air-driven pumps, sometimes referred to as 'turbo-pumps', use a concept similar to the air turbine starter but without the associated reduction gear and the pall clutch. Figure 8.9 shows a typical afterburner fuel pump arrangement.

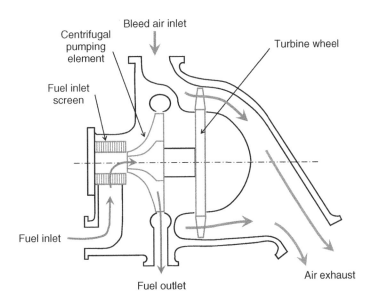

Figure 8.9   Bleed-air-driven afterburner fuel pump schematic.

Water injection pumps may require a positive displacement pumping element in order to generate the pressures required for injection onto the gas generator combustion chamber; however, the concept is essentially the same.

8.3.2   Bleed Air for Environmental Control, Pressurization and Anti-icing Systems

Bleed air for the aircraft's air-conditioning system represents a significant operating penalty to the propulsion system. In typical modern commercial aircraft this can impact fuel consumption by as much as 1–1.5%, depending upon the prevailing operating condition and the degree of air recirculation that is employed by the system.

Figure 8.10 shows a schematic diagram of a typical bleed-air system for a commercial engine application showing how two tappings are used as the bleed-air source: one from a low stage of the HP compressor and the second from a higher stage.

The bleed-air pressure required to support the air conditioning and pressurization system requirements are typically in the range 30–35 psig. During more than 90% of the operational duty cycle, this can be supplied via the lower pressure tapping. During the descent phase however, when engine throttle settings are reduced to flight idle for an extended period of time, the HP tapping takes over to provide the necessary bleed-air pressure to the various systems. As shown in the figure, this pressure source is also connected via check valves to the APU compressed air supply and to the other engine (assuming a twin-engine application). After the APU starts the first engine, this engine can therefore be used to start the second engine via the bleed-air crossfeed. A ground

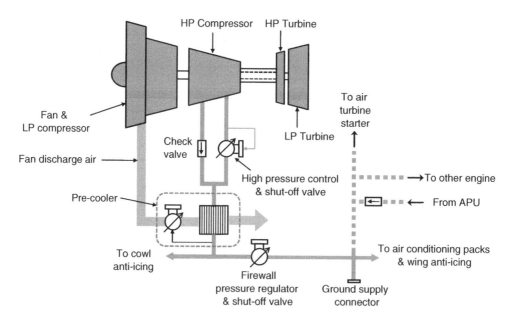

Figure 8.10   Typical commercial engine bleed-air system schematic.

air-supply cart can tap into this line to provide engine starting in the event that the APU is inoperative.

During take-off, climb, and cruise power settings, bleed-air temperatures are typically much higher than required by the various systems. For example, the low-pressure tapping can deliver air at temperatures as high as 350 °F while the HP tapping air may be as high as 1200 °F at pressures of up to 500 psig. Some form of regulation is therefore required to deliver the air from the bleed-air system to the various utilities at acceptable levels, which are typically 30 psig and 150 °F. Bleed-air condition management is accomplished using pressure regulating valves and a pre-cooler arrangement. The pre-cooler uses fan-discharge air to cool the bleed via a heat exchanger, as shown in the figure. A key system safety regulation requires that the bleed-air temperature downstream of the pre-cooler remains below the fuel auto-ignition temperature.

The nacelle cowl anti-icing system uses bleed air from the outlet of the pre-cooler, while the wing anti-icing system taps off the air-conditioning packs supply downstream of the engine firewall shut-off and pressure regulating valve.

8.3.3 Fuel Tank Inerting

In many new commercial aircraft, particularly those receiving their type certificate within the past five years, fuel tank inerting systems have been installed in order to reduce the probability of a fuel tank explosion. This is due to the loss of TWA flight 800 over Long Island Sound in 1996 which the National Transportation Safety Board (NTSB) attributed to the ignition of fuel vapors followed by an explosion within the center fuel tank. This in turn led to the introduction of a Special Regulation by the Federal Airworthiness Authority (FAA) in April of 2001. This new regulation (Special Federal Airworthiness Regulation or SFAR 88) essentially mandates inerting of the ullage within center fuel tanks, which can operate with little or no fuel and are often located close to the air conditioning packs. The heat dissipated by the air-conditioning system can contribute to the generation of high quantities of fuel vapors within these tanks, thus increasing the probably of a fuel tank explosion similar to the TWA 800 event.

While military aircraft have used fuel tank inerting systems for several decades in order to minimize fuel tank explosions due to battle damage, it was not until the last decade or so that the technology associated with 'air separation' became viable as a practical means for inerting commercial aircraft fuel tanks. The modern air-separation process uses a large number of hollow fibers assembled into a single module. These specially treated fibers have the ability to separate air into its two primary molecular components: oxygen (O_2) and nitrogen (N_2). Nitrogen, as an inert gas, can then be used to fill the ullage within the fuel tank.

Figure 8.11 shows the air separation module (ASM) concept. As shown, air enters the inlet of the air separator module and the oxygen, together with other molecules such as carbon dioxide and water vapor, migrates normal to the main flow to be vented overboard. The output from the air separator is termed nitrogen-enriched air (NEA), recognizing the fact that it is not pure nitrogen but may contain a very small percentage of oxygen (typically less than 1%).

In most cases, the air source for the ASM is bleed air which can be tapped off the same source that is used for cabin pressurization and air conditioning. The new Boeing 787

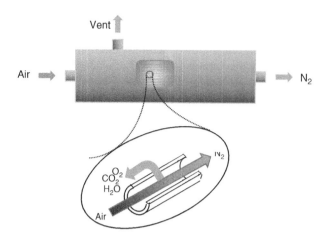

Figure 8.11   The air separation module (ASM) concept.

Dreamliner, having essentially 'bleedless' engines, uses electric motor-driven compressors to provide compressed air for all aircraft conditioning and pressurization services with the exception of the nacelle cowl anti-icing system. The latter still uses bleed air; however, this is a small demand which is used relatively infrequently.

An important issue for both the air-conditioning and fuel tank inerting systems is bleed-air contamination. From the air-conditioning system perspective, the biggest challenge is ozone contamination which is much more plentiful in the atmosphere at cruise altitudes (e.g., 36 000–43 000 ft). Ozone levels of 0.8 ppm can occur at these altitudes which can be seriously debilitating to passengers; symptoms include chest pain, fatigue, shortness of breath, headaches and eye irritation. From the fuel tank inerting perspective, ozone can seriously degrade the effectiveness of the air-separation process by eroding fiber performance and thus reducing the effective life of the fibers. To accommodate both the air-conditioning and fiber performance problems, ozone catalytic convertors are employed which dissociate ozone (O_3) into oxygen (O_2) molecules. Atmospheric dissociation also occurs when contacting ducting and interior surfaces of the air-conditioning system.

Since the ASM fuel tank inerting technology is so new to the commercial airline industry, the life expectancy of the fiber bundles contained within the ASMs is still a relatively unknown characteristic.

References

1. Moir, I. and Seabridge, A. (2008) *Aircraft Systems*, 3rd edn, John Wiley & Sons, Ltd, UK.
2. Langton, R., Clark, C., Hewitt, M. and Richards, L. (2009) *Aircraft Fuel Systems*, John Wiley & Sons, Ltd, UK.

9

Marine Propulsion Systems

All previous chapters of this book have focused on the gas turbine engine in aircraft propulsion applications. It is important to recognize, however, that the gas turbine engine has been used in both ground and marine applications to provide propulsive power, not to mention the many static gas turbine applications associated with pipelines and electrical power generation.

Of all the non-aircraft gas turbine propulsion systems, the most dominant application is the naval combatant ship where speed and agility are critical assets. Many of today's modern navies including the United States, Canada, and Great Britain employ gas turbine-powered ships, sometimes in various combinations with other propulsion engine types, to provide optimum performance over the complete operational envelope.

While gas turbines are used in merchant marine applications, they are more the exception than the rule. Similarly, gas turbine-powered ground vehicles are few and far between. Perhaps the most famous gas turbine-powered ground vehicle is the US Army's Abrams XM1 Tank which uses the AlliedSignal Lycoming (now Honeywell) AGT1500 engine; its selection as the propulsion engine is still a controversial issue with many experts. One major criticism of the use of a gas turbine for this application is the poor specific fuel consumption when operating at low power settings, due to the lower cycle temperature and hence lower thermodynamic efficiency.

In view of the above observations, this chapter focuses on naval marine propulsion and the system-related issues.

The successful development of a naval gas turbine-powered propulsion system must address a number of technical problems not common to the gas turbine propulsion systems in aircraft applications.

To put the problem into context, the typical manning requirement for a modern warship is of the order 200–275 persons. This complement of personnel can be thought of as a small village with requirements for sleeping, dining, laundry, lavatory, and so on. In addition, since the warship is a self-contained fighting machine, all facilities for combat must be maintained in an operational state with machine shops and equipment available on board. Finally, there is the requirement for mobility. The installed propulsion system must enable the ship to maneuver at low speeds in confined waters with both forward

Gas Turbine Propulsion Systems, First Edition. Bernie MacIsaac and Roy Langton.
© 2011 John Wiley & Sons, Ltd. Published 2011 by John Wiley & Sons, Ltd.

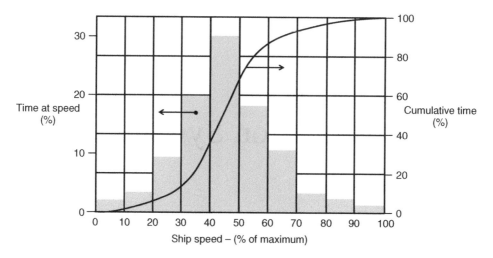

Figure 9.1   Typical warship operating profile.

and reverse thrust and to travel in open waters at speeds in excess of 35 knots. While this latter capability (which is usually classified) represents a maximum performance rating, it is nevertheless an expectation that must be addressed and met by the ship's designers.

A typical operating profile for a combative warship is shown in Figure 9.1. As shown in the figure, such a vessel sends most of its time at speeds of less than 60% of maximum. If the power requirements for such a ship with a hull tonnage in the range 10 000–12 000 tons were examined, a strong exponential relationship between power and speed would be observed as indicated in Figure 9.2.

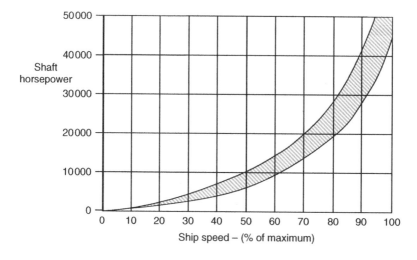

Figure 9.2   Ship propulsive power requirements.

From Figures 9.1 and 9.2 it can be seen that, for more than 80% of the time, the ship's propulsive power demand is less than 20% of the maximum installed capability. The power required of the propulsion system rises very rapidly when operating at high speeds, especially as top speed is approached.

A similar constraint applies to the electrical power generation requirements of the ship but for different reasons. While the electrical power required to support the operational needs of the ship can vary between 20 and 100% of maximum, the criticality of electrical power availability (i.e., the functional integrity of the electrical power generation system) demands than substantial redundancy be provided. As a result, the installed electrical generating capacity is usually three times (and sometimes four times) the maximum power demand.

Both the propulsion and electrical power generation systems are therefore commonly built-up from multiple sets of independent prime movers to provide both flexibility and redundancy for all of the critical war-fighting functions.

9.1 Propulsion System Designation

A series of acronyms to describe the various propulsion system configurations has been established. The following examples illustrate the concept:

- combined steam and gas turbine (COSAG);
- combined diesel or gas turbine (CODOG); and
- combined gas turbine or gas turbine (COGOG).

An example of the COGOG classification is the Canadian Navy DDH-280 Destroyer which was the first all-gas turbine-powered ship within the North Atlantic Treaty Organization (NATO) community. This ship was launched in the 1970s and its machinery layout is depicted in Figure 9.3.

As indicated in the figure, the reduction gearbox allows selection between a cruise engine rated at about 5000 HP and a boost engine rated at about 25 000 HP. Reverse thrust

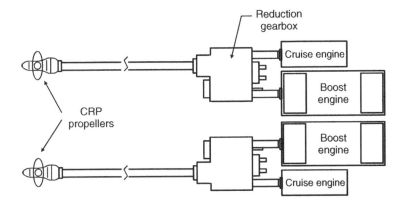

Figure 9.3 The COGOG machinery arrangement of the Canadian DDH-280 destroyer.

is obtained via a controllable reversible pitch (CRP) propeller. Finally, it is possible to shut one drive down completely, feather the propeller for minimum drag and/or noise and operate a single shaft cruise engine for either slow-speed maneuvering or economic transit.

The subject of selection of the optimum machinery arrangement for a given ship is often considered a Naval Architecture topic. However, because there are so many functional integration issues associated with the complex machinery, coupled with the operational system issues, the use of a marine system simulation is now an embedded practice in the support of the ship propulsion system design process.

9.2   The Aero-derivative Gas Turbine Engine

Many of the topics to be addressed in this chapter can be applied to any prime mover; indeed, the search for alternative propulsion systems arrangements for the warship has never really ended. The role of the aircraft gas turbine is however very important to the modern warship. The aero-derivative gas turbine will therefore be the focus of the discussion of the many issues that must be dealt with by the marine systems engineer.

The Royal Navy of Great Britain appears to have taken the lead in experimenting with and evaluating the aero gas turbine engine for marine propulsion systems [1, 2]. The United States Navy was initially of the opinion that naval requirements dictated the need for separate development of a gas turbine tailored to the marine propulsion requirement.

Considerable effort was expended in this direction including the design, development, and testing of several rather complex engines involving regeneration, intercooling, and afterburning. Interestingly, these engines were quite small with a maximum power output of less than 3000 HP; clearly, none of these engines were intended to power a ship. Rather, they were aimed at exploring the technology in a practical manner so that a degree of readiness could be achieved before committing the US Navy to a formal ship-building program with what was at the time (1940s and 1950s) a novel and very new engine concept.

In the meantime, the Royal Navy was experimenting with the aero gas turbine engine and a long and profitable association began between the US Navy and the Royal Navy. Several significant issues became apparent through these efforts.

1. The gas turbine was greatly affected by its installation and therefore inlet systems, exhaust systems, and fuel systems became separate topics of exploration.
2. Funding for separate development of marine-specific gas turbines could never approach that being used for the development of aircraft gas turbine engines.

It was about this time that the Royal Navy decided to capitalize on the large expenditures that were being made on aircraft engines and to adapt these power plants to marine applications. By the 1960s, it became apparent that a marinized aero gas turbine engine could be a candidate for the main propulsion of major warships. Several issues emerged from this concept, as described below.

1. A steam ship enjoys both forward and reverse operation through a well-developed reversing steam turbine. No such equivalent turbine existed in the world of aircraft

gas turbines nor was its development feasible. This led to the development of the CRP propeller, which was a major undertaking considering the power levels required.
2. The power levels required to achieve the speeds commonly achieved by steam-powered ships imply engines in the 30 000–50 000 HP category. Also, the fuel economy of the gas turbine at the lower power levels associated with more modest cruise speeds would be greatly compromised without the use of multiple engines. The cruise/boost engine concept therefore emerged, and is currently the norm for most naval ships.

The systems engineering problems that had to be resolved to utilize the gas turbine engine to power the modern warship were formidable. The following sections discuss the design and installation issues under several categories, the first and most obvious of which is the environment. This includes all aspects of the system which must interact with the sea or the atmosphere which is affected by the sea, for example, salt spray.

The second major area of concern is the protection of the engine itself. An airborne engine enjoys the advantage of a clean environment and a steady flow of air around the engine which helps to cool it. This is not the case for a marine engine installed low down in the hull of a ship and most certainly not the case for an engine which first found service in an airplane.

The third major area of interest is the organization and operation of auxiliary and ancillary equipment essential to its running. These include the fuel supply system, lubrication oil system, starting equipment, etc.

Finally, there are the issues of system integration and power plant control in conjunction with or in response to overall ship control. This latter issue is quite complex, since the operation of the engine must be coordinated with other engines on a multiscrew ship while at the same time maintaining surveillance over the engine from a safety and protection viewpoint.

9.3   The Marine Environment

For the purposes of developing an appreciation of the marine environment, Figures 9.4 and 9.5 are instructive. Figure 9.4 shows a merchant supply ship attempting to hold station near an offshore rig in the North Atlantic. The pitching of the ship is so severe that the stern is likely to be clear of the water. The most immediate observation is that such a condition represents a sudden loss of load on the engine. A second observation equally obvious from the photograph is the tremendous pounding of the hull structure that the ship must be capable of withstanding. It may be anticipated that the structure is designed to flex and that the ship's machinery must be installed in a manner to accommodate such movements.

Figure 9.5 is equally instructive because it emphasizes the fact that, in heavy seas, the engine inlet and exhaust are subject to conditions that may allow the egress of water. This is expected to take the form of liquid water droplets created by wind and wave interaction which are carried into the engine inlets. In the example shown in Figure 9.5, the inlet is obviously exposed to 'green' seawater which the design must be capable of accommodating.

Operationally, the marine environment demands longer-duration running of the propulsion system than the typical aircraft mission. In peacetime, warships will typically spend six weeks at sea before returning to port; wartime assignments can be several months.

Figure 9.4 Offshore rig supply vessel with stern out of the water (courtesy of lyman@naval.com).

Figure 9.5 Coastguard vessel operating in heavy seas (courtesy of lyman@naval.com).

As was noted previously in Figure 9.1, much of the time the ship's demand on the propulsion system are relatively modest; however, they must continue to run and be available as needed for the prescribed operation.

9.3.1 Marine Propulsion Inlets

While Chapter 6 of this book addresses gas turbine inlet and exhaust systems, it is confined to typical airborne applications. Since the inlet and exhaust systems of marine applications are substantially different, having to deal with the unique installation and environmental

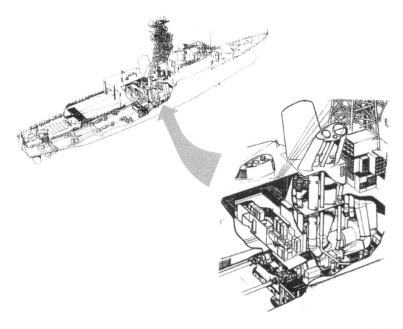

Figure 9.6   DDH-280 propulsion machinery arrangement (courtesy Lt. Cdr Taylor).

considerations, marine inlet and exhaust systems for marine applications are addressed separately in this chapter.

The marine propulsion inlet system is arranged to duct air from a point high up in the ship's superstructure down through several decks to the engine, which is typically located not much above the point where the propeller shaft protrudes from the hull. Figure 9.6 provides a cut-away view of the propulsion system installation of the Canadian DDH-280 Class Destroyer, indicating the shafting arrangements, the location of the engine enclosures, and the arrangement of the inlet and exhaust systems.

If we compare the COGOG schematic shown in Figure 9.3 with the actual arrangement shown in Figure 9.6, it can be seen that there is separate ducting for the cruise engine and the boost engine. It is also apparent from Figure 9.6 that the inlet air must be drawn down and around a 90° bend before it is presented to the engine.

Shepherd [3] describes flow through a bend in a manner which exemplifies the problem. As the air flow enters the bend in a pipe, the curvature of the axis causes the pressure on the outside radius of the pipe to increase due to centrifugal effects. Similarly, the pressure on the inside radius will decrease.

As the flow passes the midpoint of the curve, the effect is reversed. This effect, combined with fluid friction, sets up twin circulation patterns around the periphery of the duct. If the velocity is high enough, separation will occur on the inside diameter of the bend. These two effects combine to produce a highly distorted flow at the outlet of the bend. Furthermore, in cramped spaces typical of warships, there is no space to allow the flow to straighten itself before entering the engine.

Under such inlet flow conditions, portions of the engine inlet annulus can be starved of air leading to a compressor stall and/or rotating stall. It is therefore necessary to make the ducting as large as possible, in order to slow down the air flow rate as much as possible and induce the flow to traverse through the bend through turning vanes prior to final acceleration to the engine face.

Obviously, the above commentary relates to the design point or maximum flow conditions. At part power conditions, the airflow demands of the engine are lessened and the flow will be somewhat more benign; however, when under maximum power conditions it is important that the losses in the inlet do not reduce the capacity of the engine which, in turn, would compromise the capability of the ship. A careful aerodynamic design of the inlet ducting is therefore warranted.

9.3.1.1   Inlet Treatment

The principal concern in the treatment of marine inlets is the inevitable presence of water in the form of mist, water droplets and, in the worst case, so-called green water washing over the mouth of the inlet. In short, the inlet system must be designed to deal with water in various states in a manner that will allow the engines to function continuously. To this purpose, the inlet is fitted with water separators, de-misters, and filters.

The first defense against water ingress is a system of louvers arranged as indicated in Figure 9.7. The combination of placing the inlet high up in the ship's superstructure and arranging it to shed green water removes much of that problem; however, water borne by the air in the form of spray and mist will be carried into the ducting and must be separated. In this case, coalescers are used to reform any fine mist into water droplets which can then be easily removed by gravity.

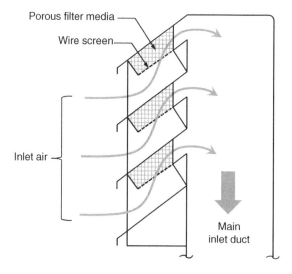

Figure 9.7   First stage inlet water removal.

Downstream of the coalescers, the inlet is fitted with additional moisture separators designed to deal with residual moisture in the air as well as any remaining water droplets. Typically, the air is passed over a system of louvers which presents a substantial surface area to encourage further formation of water droplets. These louvers must be heated (usually by electric heaters) to prevent ice formation. Finally, the air must pass through a series of inlet screens which facilitate the removal of particulate and any remaining small water droplets.

Several other features comprise a typical marine inlet system as follows. Firstly, there is very real concern over the build-up of ice which can cause blockage and subsequent engine compressor stall. In the event that ice does accumulate, there is a further concern that the ice can cause foreign object damage (FOD) to the engine. To mitigate this problem, it is common practice to bleed air off the engine compressor and to route this hot air to an anti-icing manifold in the inlet. Sensors are provided in the inlet to indicate the increased risk of icing as ambient air falls to near freezing (about $+3\,°C$), at which point the anti-icing system is turned on.

Secondly, the inlet is fitted with blow-in doors that will open if the pressure drop across the inlet rises above a certain value. Such conditions indicate that the engine is being starved of air due to a blockage within the air treatment system.

Once the blow-in doors are opened, the only protection that the engine enjoys is the final inlet screen to block foreign objects. It is expected that continued operation under such conditions will result in salt-related damage. As a minimum, therefore, the engine compressor must be washed following such an event.

Finally, the inlet is treated with acoustic material within the ship (normally about half of the total inlet ducting). This is needed to suppress any aerodynamic noise generated within the inlet ducting. This treatment usually takes the form of vertical louvers of sound-absorbing material covered by a skin of perforated steel plate. The perforations are tuned to attenuate the dominant noise frequencies.

The inlet system is terminated at the engine enclosure. This final component is a rubber-boot-like piece of ducting designed to guide the air into the engine enclosure without transmitting enclosure noise to the ship's hull.

9.3.2 Marine Exhaust Systems

Exhaust gases from the engine must be routed back up through the ship, again to a point high up in the infrastructure. As with the inlet ducting, the exhaust ducting is close-coupled to the engine and must initially execute a right-angle bend before traveling up through the ship.

It is worth noting at this point that the engine is operating within an enclosure and that it requires cooling air in much the same manner as is required for aircraft applications. For this reason, air is drawn in separately from the inlet and is pumped into the enclosure and around the engine. This air is then drawn out of the enclosure by an arrangement of eductors around the periphery of the power turbine, energized by the hot high-speed exhaust gas. This air helps to cool the exhaust gas prior to executing the $90°$ bend before the uptake to final exhaust.

Throughout the entire piping of the exhaust system, some form of acoustic treatment is required. It should be emphasized that the exhaust gas temperature of a typical high-performance marine gas turbine will be in the vicinity of 500 °C (930 °F). Such a temperature is close to the upper operational limit of many conventional steels, so a secondary liner may be required in most applications. Turning vanes at the first bend are an absolute necessity.

Perhaps the most significant feature of the exhaust duct is the infrared suppression system intended to minimize the possibility of heat-seeking missiles acquiring the ship as a target and, for that matter, to prevent any airborne surveillance system being able to 'look down exhaust stack' of a modern combatant warship. Infrared suppression systems are usually fitted to the top of the exhaust stack, and operate to cool the exhaust plume as much as possible and to prevent line-of-sight access to the stack.

A generalized arrangement of a typical infrared suppression system is shown in Figure 9.8. The system shown consists of a ball-shaped obstruction to the exhaust gas which completely blocks any possible direct line of sight down the stack. The ball is fitted to a larger duct which, when installed on the top of the exhaust stack, forms an eductor which draws cooling air into the system. Because the ball also redistributes the exhaust into an annulus, the cold air mixes with the exhaust thereby reducing its infrared emissions.

A final comment on the exhaust ducting is that the possibility of exhaust gas being inducted into the engine inlet must be avoided. This is achieved by ensuring that the exhaust funnels are higher and well aft of the inlet system.

9.3.3 Marine Propellers

Ship propulsion has traditionally been achieved through the use of propellers. During the early steam era, these propellers were single-piece affairs driven by a steam turbine fitted with forward and reverse turbine wheels, with valving arranged to allow steam to be directed to one or the other. Even the high-power diesel marine propulsion systems used simple propellers driven via a reversible gearbox and clutch mechanism. These systems were simple, easily controlled, and, more importantly, provided rapid response.

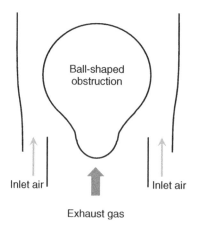

Figure 9.8 Typical infrared suppression system.

Once it was realized that a reversing power turbine fitted to an aero-derivative gas turbine engine was not a realistic possibility, the CRP propeller (sometimes referred to as a controllable pitch propeller or CPP) was developed to be operated with the gas turbine engine.

This type of propeller, as indicated in Figures 9.9 and 9.10, is a relatively complex device. Each blade is arranged to turn about a spindle within the hub, thereby providing variable pitch capability. The pitch-changing mechanism is embedded in the propeller hub; however, the actuating mechanism may be located within the hub as shown in Figure 9.9 or remotely inside the ship as depicted by Figure 9.10.

While the pitch-control changing mechanism is an integral part of the CRP system, propeller pitch commands must come from the overall ship propulsion control system.

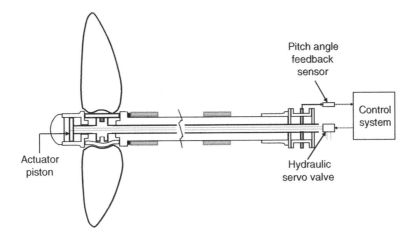

Figure 9.9   Typical CRP with the pitch actuator located in the hub.

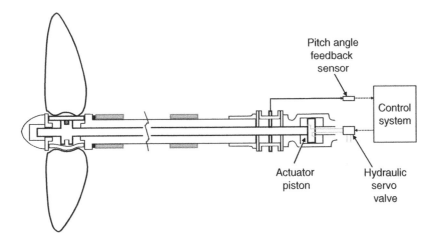

Figure 9.10   CRP propeller with the pitch actuator remotely located.

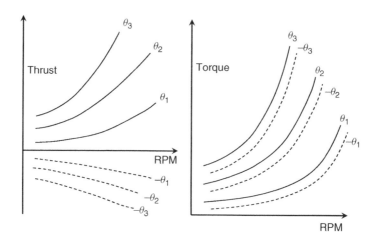

Figure 9.11   Typical CRP propeller characteristics.

Here, the control of propeller pitch is carried out in conjunction with the prevailing engine power and ship speed; a measure of propeller pitch angle is therefore a key input parameter to the system controller.

Apart from reverse thrust capability, the principle advantage of the CRP is the ability to set the pitch angle to match the power requirements of the ship at various ship speeds. Figure 9.11 shows how thrust and torque varies with shaft revolutions per minute and propeller pitch for a constant hull speed.

One interesting feature of the CRP system is that several combinations of shaft speed and propeller pitch can develop similar values of thrust, but at different torque demands from the propulsion system.

This issue is best settled by a comprehensive dynamic analysis of the ship and its machinery over the complete operating range. From these analyses, it then becomes possible to evolve a control strategy that is appropriate to the design at hand.

Since most warships have multiple engines (COGOG, CODAG, etc.) it is reasonable to assume that there will be a number of different machinery modes of operation, each of which will require some exploration of possible control system settings for a given ship speed.

It should also be mentioned that the hull resistance is dependent on the weight (draft) of the ship and the general hydrodynamic condition of the hull. Similarly, the overall performance of the propeller is dependent on its condition.

9.4   The Engine Enclosure

Although not strictly necessary in the context of overall ship propulsion system design, the industry has judged it convenient to locate the gas turbine in a self-contained enclosure. This arrangement allows each engine supplier to 'pre-engineer' the package, thereby resolving most of the major installation issues pertaining to the engine. A schematic of the enclosure for the GE Marine LM2500 is shown in Figure 9.12 [4].

The primary goal of this design is to protect the engine and to provide a straightforward package for the shipyard to deal with. In this application, parts of the lubrication system and integrated electronic engine controller are remotely located.

In other marine propulsion packages, for example, the Rolls-Royce Marine Spey SM1A [5], the plant control and the significant functions associated with the fuel control and lubrication systems are part of the enclosure assembly.

9.4.1 The Engine Support System

There are several issues to be considered in the mounting of the gas turbine in a ship. This is particularly true of warships, and includes:

- shaft alignment with the marine drive;
- shock and vibration; and
- provision for cooling.

As can be seen from Figure 9.12, the gas generator is mounted centrally to the enclosure walls. The power turbine is bigger in diameter and must be firmly mounted on the enclosure structure in order to react to the forces and moments associated with its operation. The gas generator is subsequently mounted to it and supported front and back by flexible, adjustable supports. A flexure is shown in the front gas generator mount to accommodate thermal growth of the engine in operation.

The second major feature of the engine enclosure is that it is mounted on a 'raft' on the hull superstructure. The enclosure itself is made quite stiff; however, this raft-like structure is mounted on shock and vibration fixtures relative to the ship's hull. Such an arrangement tends to dampen out any engine-generated noise and vibration that might otherwise be transmitted to the ship's hull.

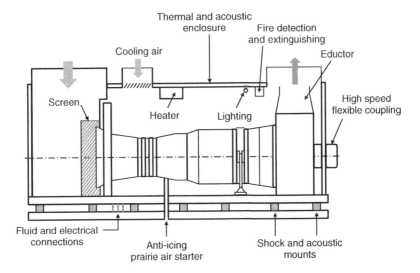

Figure 9.12   Engine enclosure module for the GE Marine LM2500.

In any case, the overall installation must survive the type of shock that can occur from the detonation of a torpedo in the water within a close proximity to the hull. The shock mounts are designed to absorb most of the energy from such a blast and while the engine may shut down in certain circumstances, it is expected to survive and remain operational.

All gas turbine installations for naval and marine applications deliver power to the propeller through a large reduction gearbox mounted directly on the hull structure. This suggests some relative motion between the gas turbine and the gearbox, which is accommodated via a high-speed flexible coupling as depicted in Figure 9.12.

9.4.2 Enclosure Air Handling

The typical engine enclosure provides a large duct connection for the air supplied to the engine. This is delivered in the case of the Rolls Royce Spey SM1A through a system of cascade turning vanes before entering the engine. On the other hand, the GE Marine LM2500 incorporates an inlet plenum that is much larger relative to the engine airflow demand and therefore turning vanes are not employed. A final inlet screen covering the mouth of the engine air inlet is used in the GE application to protect the engine from possible FOD.

The enclosure also provides a barrier preventing inlet air from entering the engine compartment and a separate air inlet is provided to duct cooling air into and around the engine. This air is induced to flow into the enclosure by an eductor arranged between the engine exhaust and the ventilating and cooling air flow as shown in Figure 9.12. This simple arrangement for cooling the enclosure generates a substantial vacuum relative to the ambient conditions, and the doors to the enclosure are securely sealed to prevent leakage.

As noted previously, the exhaust is mounted to the ship up-take ducting using a short rubber boot which ensures a seal and, at the same time, allows for movement of the enclosure relative to the intake.

9.4.3 Enclosure Protection

The entire engine enclosure is clad in acoustic material which provides both noise suppression and thermal insulation.

In addition to the lighting systems, the enclosure is fitted with a fire-protection system. This consists of smoke and flame detection as well as a fire-extinguishing release system employing either CO_2 or Halon. In most modern installations, this system is monitored through the overall ship's control system and may be initiated either manually or automatically from the central control.

The final element of the engine enclosure is the provision of an engine water-wash system. This is a basic requirement for a gas turbine in a marine environment. The basic elements of an engine water-wash system are:

- a supply of fresh water;
- a source of detergent;
- a pump or pressurizing tank which permits mixing of water and detergent; and
- a means of delivering the wash fluid to the inlet bell-mouth of the engine.

Operation of the water-wash system involves motoring the target engine on its starter and spraying the wash liquid into the mouth of the compressor. Typical systems include the ability to spray only fresh water in order to rinse the engine before it is allowed to drain. The water-wash operation is conducted periodically (e.g., once per month) or after an event such as a breach of the normal inlet protection system which is usually associated with filter/demister bypass.

9.5 Engine Ancillary Equipment

Ancillary equipment comprises those systems specifically required to run an engine in the ship environment, including:

- lubrication oil system;
- fuel control system; and
- engine starting system.

Since most gas turbines in marine service are aero-derivatives types, the basic on-engine components of these systems are unchanged. The systems required to accommodate marine applications, however, must be compatible with the engine in question. They are therefore either specified by the engine supplier or provided to the ship as part of the engine package.

9.5.1 Engine Starting System

The basic requirements of a starting system for marine gas turbines is a means of cranking the engine up to light-off speed which, in all respects, is similar to that required for aircraft engines. However, the requirement to motor the engine for purposes of washing and rinsing extends the demands on the starter system.

Typically, the choice of energy source is pneumatics since compressed air is readily available from a variety of sources such as ship-board compressors, stored compressed air, auxiliary power units, and bleed air from another operating engine. The majority of naval warships are fitted with at least two gas turbines; compressed air from the gas generator compressors is used for a number of purposes including engine inlet anti-icing, noise masking (through bubbling of air out through the hull), and for starting other ship-board equipment.

Such practices decrease the available power from the engine from which air is extracted and, as a result, increases its fuel consumption; however, the convenience and availability of compressed air makes its use attractive.

A typical air starting system is shown in block diagram form in Figure 9.13. If we consider the bleed air from another gas turbine as a primary source of compressed air, note that this air will be extracted from the engine at thermodynamic conditions of some 20 atm and at temperatures of the order 800–900 °F (420–480 °C). By contrast, a typical air starter requires air at 3–6 atm and at temperatures of the order 100 °F (40 °C). Cooling of bleed air down to temperatures into the range suitable for the air starter is achieved most readily by the use of seawater.

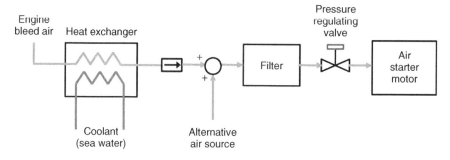

Figure 9.13 Typical pneumatic starting system.

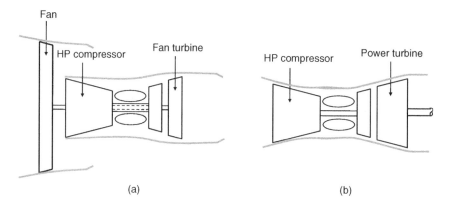

Figure 9.14 Evolution of the aero-derivative gas turbine.

The reliability of the air starter motors is inevitably related to the quality of the air used. For this purpose, air filtration in the form of either inertial separators or a barrier filters are used to protect the motor.

Finally, an automatic pressure regulator is fitted to drop the pressure of the air supply down to a level appropriate for the pneumatic starter motor. By regulating the air pressure, it is possible to control the speed of the starter motor to match the prevailing mode of operation, that is, engine starting or water-washing and rinsing.

Also shown in Figure 9.13 is the provision to automatically select an alternative source of compressed air. In many installations, provision is made for storage of high-pressure air via a low-volume compressor system to levels of 250–275 atm. The air is dried prior to storage in bottles designed to handle pressures at this level. Once activated, these bottlers are 'blown down' through a two-stage pressure regulator from 275 to 40–50 atm and from there to the level required for the starter system. There are no set rules for the capacity of such a system; however, it is common practice to make provision for 4–6 starts using such a source of compressed air.

It should be noted that this system, as with any blow-down air supply system, will suffer from a total temperature drop in accordance with the laws of thermodynamics.

Assuming an adiabatic expansion from P_1 to P_2, the temperature ratio is given by:

$$\frac{T_2}{T_1} = \left(\frac{P_2}{P_1}\right)^{\frac{\gamma-1}{\gamma}} \tag{9.1}$$

where γ is the ratio of specific heats c_P/c_V (=1.4 for air).

An expansion from 275 to 200 atm (say, one start) therefore yields a temperature ratio of 1.095 which translates to a fall in temperature of about 50 °F (22.5 °C).

If the air is not very dry, moisture will begin to precipitate as ice crystals which can be a major problem for the air starter motor. Therefore, either the air is kept very dry or some means of reducing the temperature drop must be sought.

In blow-down wind tunnels, this is achieved by passing the air through a large thermal mass or heated plates. In the case at hand, it is possible to start a single engine on the first step and then mix the two streams of air (one from the compressor bleed of the operating engine and the other from the blow-down tanks). This practice will lower the temperature of the bleed air and, at the same time, raise the temperature of the blow-down air.

9.5.2 Engine Lubrication System

The aero-derivative gas turbine in marine applications retains many of the components of the lubrication system common to its airborne parent.

Most modern aero engines adapted to marine applications began life as turbofan engines. For marine shaft power applications, the fan is removed and its driving turbine is redesigned to operate as a power turbine as indicated in Figure 9.14.

Since elements of the lubrication system such as oil sumps, coolers, de-aerators, etc. are usually designed to be attached to the outside of the fan casing. All of these components are redesigned as separate units which are part of the engine functionally, but are either completely separate from the engine enclosure (e.g., the GE Marine LM2500) or integrated into the engine enclosure module as a separate package (e.g., the Rolls Royce Marine Spey).

A functional block diagram of a lubrication system for a typical marine application is shown in Figure 9.15 which depicts how the lubrication supply package is installed off the engine. It is common practice in marine applications for the lubrication supply system to be mounted above the engine with an elevation difference of the order 10 ft. This arrangement facilitates a gravity feed to the engine, which provides a continuous supply of oil for a short period during engine shutdown.

As many of the aero-engine components as is practical are retained in marine service applications. The oil supply pump and the multistage scavenge pumps are therefore driven off the accessory drive gearbox which, in turn, is driven by the gas generator shaft. It is through this accessory gearbox that the starting power is applied to the engine. Similarly, the high-pressure fuel pump is driven off this gearbox.

All of the lubrication monitoring functions (filter pressure drop, oil pressure, and temperature) are retained from the aero version of the engine. Some piping changes are necessary to accommodate the addition of oil debris monitors both upstream and downstream of the scavenge pump.

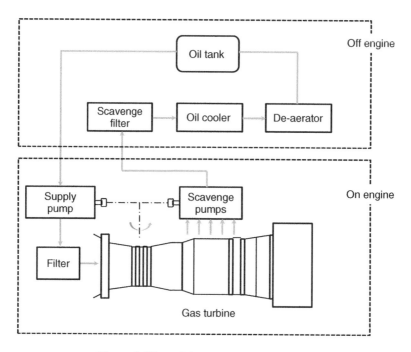

Figure 9.15   Lubrication system functions.

The off-engine lubrication supply system is also fitted with monitoring equipment, including:

- scavenge filter pressure differential transducer;
- oil temperature sensor;
- oil level indicator; and
- coolant temperature sensor.

In the modern warship, all of these system health-monitoring parameters are brought to a local control panel as well as being made available to a ship-wide control and monitoring system.

9.5.3   Fuel Supply System

The supply of fuel to a marine gas turbine assumes that the engine requires a clean distillate similar in most respects to the fuels used by its aero-engine parent. The principle problem facing the marine operator is the ever presence of water (both salt water and fresh water) in the fuel. The shipboard fuel supply system is therefore largely concerned with the removal of water from the fuel before it is delivered to the engine. A typical functional block diagram is shown in Figure 9.16.

Fuel is stored in several large tanks distributed throughout the ship at low points within its structure. These fuel tanks are an important component of the ship's ballast system,

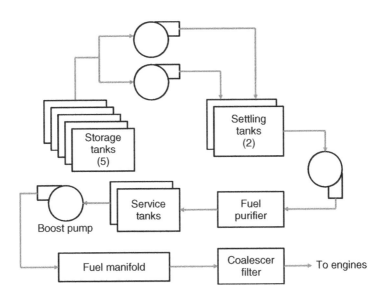

Figure 9.16   Typical ship fuel supply system.

and it is common practice to replace fuel with sea water to ensure proper distribution of weight.

It is now well known that the interface between fuel and water promotes the growth of microbes which, without proper treatment, find their way through the fuel supply system to the coalescers rendering them ineffective. Modern practice favors the use of bladders or other physical barriers between the fuel and the ballast water in the fuel storage tanks.

Referring to Figure 9.16, redundant pumps transfer fuel from the main storage tanks to settling tanks, whose principle function is the separation by gravity of particulate and water. From the settling tank, the fuel is pumped to a fuel purifier which is basically a centrifuge comprising a stack of conical disks [6].

Typically, fuel enters the purifier through a central port or core and passes through a rotating disk set from the outside perimeter to the inside. The disk stack is set up with allowance for thin flow passages. As the fuel passes through the disk passages, centrifugal force separates any particulate matter and water from the fuel. By arranging for a seal between the underside of the top disk and the body of the centrifuge, the water is contained on the outside of the disk pack and the fuel is confined to the inside. Water can therefore be drawn off as waste and the remaining purified fuel is delivered to the service tanks as shown in the figure.

From the service tank, the boost pump delivers fuel to the fuel manifold and to a final treatment by the coalescer filter [6]. The coalescer consists of a fiber matt filter which separates solid particulate down to 5 μm, while agglomerating water into droplets which are then removed from the filter under the action of gravity. In essence, the coalescer is a filter providing a barrier which preferentially passes fuel while stopping and/or agglomerating water. In the event that microbes grown in the fuel storage tanks reach the coalescers, they will most certainly contaminate the coalescer filter element necessitating its replacement.

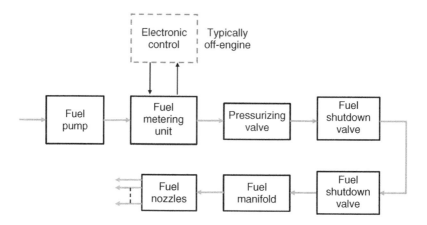

Figure 9.17   On-engine fuel system functional diagram.

In some coalescers, a final polytetrafluoroethylene- (PTFE-) coated stainless steel screen is provided which further assists in the trapping of liquid water.

From the coalescer, the fuel is delivered to the engine fuel delivery and control system which is, for the most part, identical to the airborne version of the engine. A top-level block diagram of the on-engine fuel delivery system is shown in Figure 9.17.

The primary addition to the marine fuel system shown in the figure is the incorporation of redundant fuel shutdown valves. These valves represent the final protection of the engine and the ship from possible fire and/or explosion due to an overspeed, false start or any other emergency condition that demands an immediate shutdown of the engine. Typically, these valves are closed in a non-energized state and will snap shut immediately upon removal of the enable command.

In all other respects, the fuel system retains the functional features of the airborne version of the engine including acceleration/deceleration limiting, variable compressor geometry control, bleed valves and protection features, etc. that are essential to the safe operation of the engine. This equipment is supplied as part of the marine engine package.

9.6   Marine Propulsion Control

9.6.1   Ship Operations

No discussion of ship machinery control can be undertaken without some awareness of the operations that may be demanded of the ship by its commanding officer (CO). These may be usefully divided into steady-state and transient maneuvers. From the CO's perspective, he wants the capacity to propel the ship forward at any speed from 'dead slow' to 'full ahead'. In addition, he wants the ship to be capable of traveling in the astern direction although 'full astern' would be considered an unusual request insofar as a steady-state condition is concerned.

Transient maneuvers are usefully divided into what might be classed as normal changes from one speed to another and 'crash' changes where the CO is requesting ship speed changes with as much urgency as can be obtained from the propulsion machinery. In all cases, the performance of the ship is determined by its response to these demands. This is a very complex systems issue, involving the stability and resistance of the hull to changes in speed and direction as well as the various modes of operation of the machinery and its control. Specifically, changes of such magnitude create complex interactions between the ship's wake and the propeller. Such flow regimes profoundly affect the behavior of the propeller and consequently its capacity to respond to the demanded changes in speed.

In steady-state operations, the response of the ship is determined by the propeller fitted to it and the range of pitch and rotational speeds of which it is capable. Let us consider the steady operations as a starting point. At zero speed (calm waters), we can safely assume that a stopped shaft with propeller pitch set to zero will be associated with such a condition. Acceleration to any forward speed requires some combination of propeller shaft speed and pitch to achieve this new state. It is important for the range and endurance of the ship to determine the optimum combination of shaft speed and pitch with respect to power absorbed. Thus, from a design perspective, it is possible to compute the optimum settings of pitch and shaft speed for a given steady-state speed demand of the ship. Such a calculation requires knowledge of the hull hydrodynamic characteristics and the propeller characteristics.

Having made such a calculation, we then proceed to the machinery. Here, the designer is presented with a number of choices, of which the following are typical.

1. Can this ship speed be provided within the power range of the cruise engine?
2. If not, can it be handled by the boost engine?
3. Once the engine(s) have been selected, what combination of power settings is best suited to the demanded condition?

The above questions imply a particular machinery arrangement (i.e., engines, gearboxes) from which the choices can be made. Clearly, the choice will depend upon the overall efficiency of the power delivered in the context of fuel consumption. While the previous calculation may have provided an optimum choice of propeller pitch and shaft speed, from a power demand perspective the conversion of that demand to an engine setting which results in minimum fuel consumption may dictate another set of choices.

The number of combinations of machinery arrangements possible through compounding gearboxes, allowing more than one engine to contribute to the power requirements of a specific propeller, argues strongly for the use of computer simulations as a means of selecting the most sensible combination of prime movers, gearbox arrangements, power settings per engine, and propeller pitch settings for each ship speed. Such a tool allows for a rational evaluation of the choice of machinery, followed by a selection and evaluation of control algorithms leading to a high-quality design.

The above discussion only considers the steady-state operation of the ship. It establishes, for a given ship's speed demand, a choice of machinery arrangement; together with pitch and throttle settings of that suite of machinery, the required ship's speed is achieved. This highlights that a given machinery arrangement cannot provide the full range of ship's speed that is possible or, if it can, there will be parts of such an operating envelope which are so inefficient as to prohibit their use. It is therefore to be expected that a transient maneuver from one steady-state condition to another may well involve a change of machinery modes. Such changes in the selection of engines and/or gearbox clutches represent substantial complexity of control if they are to be done 'on the fly'.

In terms of automation, the marine/naval community has been relatively slow to adopt new technologies compared to other industrial applications of the gas turbine. For example, it is perhaps 40 years or more since pipelines have been starting, stopping and operating a mechanical drive gas turbine from a remote location; no manual operation is ever considered in this industry today.

By contrast, a warship will spend substantial amounts of time at sea with few, if any, opportunities to enter port. From an operational perspective, a warship is therefore an independent operating unit. It must be designed such that the crew can deal with virtually any operational eventuality, including battle damage. It therefore follows that any machinery control must have the capability to function manually. In an engineering sense, it should be possible to automate all mechanical systems such that the ship can operate from a single command center. It can be brought to any operational state with a single command. It can self-diagnose faults of any piece of equipment and, to the extent physically possible, provide a way to work around these faults. To the propulsion control systems specialist, a ship can be arranged such that it is completely unmanned. As a consequence, it is technically possible for several ships to be commanded from a central or lead combatant.

The above arguments have been made at ship's control system conferences for at least the past three decades; however, the practical sailor or ship commander has so far resisted a rush to automation with a number of arguments as follows.

1. A marine system is much too complicated for any practical person to believe that technology can provide a complete answer. For this reason alone, it is imperative that all ship's systems, including propulsion, can be isolated from the overall ship's control system and operated independently both as an automatic system and as a manually operated entity.
2. A warship is likely to sustain battle damage. It cannot be predicted what functional elements will be damaged. As a consequence, human ingenuity is critical in such situations. Complete freedom to take over the equipment may be essential to the survival of the ship and its personnel.
3. A warship needs to be manned by personnel with sufficient training to be able to handle any situation. Such training can only be obtained by 'hands-on' operation of the ship and its systems.
4. The ship is a fighting unit which is commanded from the bridge. The complexity of an engagement demands that the CO be able to focus on the war fighting elements.

The above arguments, together with the relative infrequency of ship development programs, have resulted in slow but very deliberate progress in the area of ship's control. To put this in perspective, a warship powered by steam typical of the 1960s had approximately 150 parameters that were fitted with sensors and monitored either continuously by an elementary control or by a human operator. Canada launched the first all-gas-turbine warship in 1970 (COGOG arrangement). It was fitted with a hybrid pneumatic, analog electronic control system with 750 parameters either monitored or controlled. In 1990 Canada also launched the first warship with all-digital control and a centrally managed suite of ship's machinery (CODOG arrangement). This system has a central bus with remote terminal units at various locations on the ship. Over 5000 ship's machinery parameters are either under surveillance or under direct control. Nevertheless, all of the above systems could be operated in isolation and in manual mode.

In practical terms, the progress in marine propulsion controls has been primarily in the field of improved systems operability. There has certainly been some reduction in crew workload; however, the major progress has been made in the capacity to monitor the behavior of the machinery and to present this greatly enlarged array of data at a man-machine interface in a manner that facilitates rapid assimilation by the officer of the watch.

A full description of a complete ship's automation package is beyond the scope of this book; however, the following description of the propulsion system control will give the reader a feel for the overall system.

9.6.2 Overall Propulsion Control

An overview of a typical ship's propulsion control is shown in Figure 9.18. There are several factors which influence the organization of such controls. First, an entire propulsion system is an assembly of components that are highly ship-specific. A hull of a specific size is defined by the Naval Architect in response to an operations mandate stated by experienced Naval Combat Officers. The speed and maneuverability of the ship will, in turn, dictate the size, shape, and speed requirements of the propeller(s). Without exception, the rotational speed of the ship's propeller(s) is substantially lower than the typical rotational speed of a power turbine. This defines the specification for the reduction gearing, which must match both the engine and the propeller.

Currently, for reasons discussed previously, the aero-derivative gas turbine and CPP remain the most cost-effective combination for a warship. This in turn drives the requirement for integrated control of major pieces of equipment supplied by several sources.

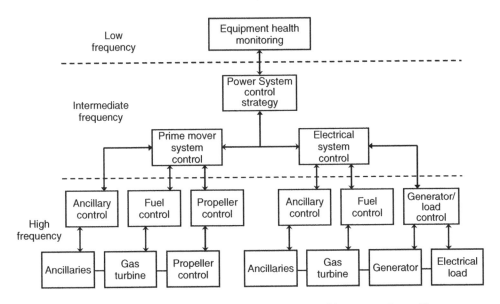

Figure 9.18 Typical propulsion control for a gas turbine-powered warship.

The second major factor which is partly business-driven and partly technical is the desire of component suppliers to specify the control features of their unit. In the case of the gas turbine which has the possibility to surge the compressor, overspeed the gas generator, produce over-temperatures, and other critical conditions, the arguments are largely technical. Such unit controllers are highly specialized; the gas turbine supplier who must provide guarantees on performance and reliability of his machine rightfully demands that it be controlled in accordance with his specification. The provision of a unit controller by the engine supplier is the surest way to protect the engine from damage caused by inattention to details, either in ship design or in operation.

Figure 9.18 indicates the parsing of control functions based on frequency response. It also implies multiple processors with primary responsibility for accurate and stable control of the unit to which it is dedicated, with a secondary role to communicate to the other processors above it in the hierarchy. Such an argument is borne of the notions of multiple processors sharing a common bus through which information is passed. Although most processors today are sufficiently fast and low enough in cost to handle the workload, the division by frequency response remains useful. It allows for the intervention of a human being at local operating panels if, and when, there is a need to assume manual control of the machinery in question.

From an algorithm or logic perspective, the entire propulsion control comprises three principal functions:

• propulsion system monitor (PSM);
• propulsion system controller (PSC); and
• propulsion system sequencer (PSS).

The propulsion system monitor processes all measurement points within the system and, from these data, the state of the prime mover and the propulsion drive train is determined. These state variables are communicated to both the PSC and the PSS in response to either commands from the bridge or changes in the state of the machinery. It is this component of control that provides all annunciations and display of operating data.

The PSC is that component of the overall control system which maintains continuous control of the machinery currently in operation. This includes scheduling of propulsion machinery demands for propeller pitch and turbine gas generator speed, and for verifying that this machinery does not exceed steady-state or transient limitations. Note that unit controllers typically supplied by the manufacturer are considered to be an integral part of the PSC. For example, when the PSC responds to a demand for increased ship speed, it schedules an increase in power which is then converted to a demand for change in the gas generator speed. This latter demand is met by the gas turbine unit controller supplied with the engine.

The PSS is that part of the overall control which processes the drive mode change over requests. This is principally a logic processor since actual unit control is handled by the PSC. However, the propulsion system sequencing function handles the putting of prime movers into and out of drive mode, the synchronization of shaft speeds during changeovers, and the controlling of load application rates.

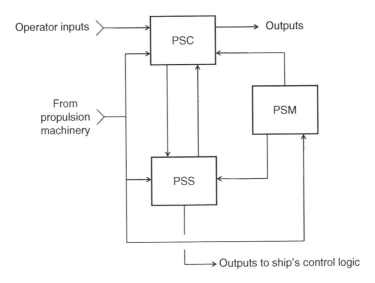

Figure 9.19   Overview of the ship's propulsion control logic.

This control system arrangement implies multiple processors with primary responsibility for accurate stable control of the unit to which it is dedicated, with a secondary role to communicate during these changes. A simplified overall propulsion logic diagram is shown in Figure 9.19.

As shown in the figure, the propulsion system monitor maintains surveillance over machinery parameters in order to determine drive mode indicators, gas turbine state indicators, etc. Also indicated, the PSC receives inputs from the operator and from the machinery. These inputs include an operator (or bridge) command for a specific ship speed. In the event that the speed demand exceeds the capability of the current drive mode, the operator may request a change of prime movers.

The function of the PSS is to execute the drive mode changeovers as requested from the PSC. During such a changeover request, the sequencer takes over control from the PSC in order to synchronize shaft speeds and control application of loads during the changeover. Once the changeover is completed, the sequencer relinquishes control to the PSC after which normal response to operator demands for a specific ship speed is resumed.

9.6.3   Propulsion System Monitoring

The propulsion system monitoring function provides surveillance over the entire propulsion machinery train. These data are gathered and presented to the operator via a graphical user interface with specific annunciation of the machinery drive modes and the value of individual parameters presented in comparison to limits. A photograph of a control center together with a typical 'page' from these computer screen-based control panels is shown in Figure 9.20.

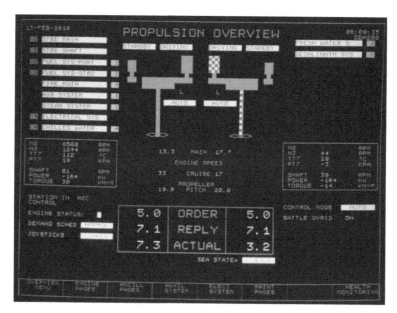

Figure 9.20 Control center and typical graphical user interface presentation of propulsion system state data (courtesy of GasTOPS Ltd and DND Canada).

In this type of presentation, use is made of color to indicate warnings and alarms in the event that such a situation would occur.

In addition to presentation of state data, several indicators are passed to the propulsion controller and to the PSS as follows:

- boost engine state indicator;
- cruise engine state indicator;
- shaft state indicator;
- drive mode indicator; and
- drive mode schedule indicator.

The cruise and boost engines will typically have similar states as described by Table 9.1.

It should be noted that the STANDBY state indicates a higher state of readiness than the OFF state. This implies that the ancillaries are operating and that the engine is ready to start.

The shaft state is determined to be either LOCKED or NOT LOCKED. In the locked state, the shaft is clutched such that it is not free to rotate. In the NOT LOCKED state, the shaft is free to rotate and is said to be 'trailing'. In such a condition, the synchronizing self-shifting (SSS) clutch allows the shaft to rotate freely if power is not applied. As power is applied, the clutch engages and the propeller follows the propulsion system input.

The drive mode indicator is derived from other individual state variables and is intended to establish the configuration of the machinery on each shaft line. For example, in a COGAG arrangement, both engines can be driving. In this case, the DRIVE mode might be declared as 'COMBINED' with the following specific state parameters required for its determination:

- boost engine on-line;
- boost engine clutch engaged;
- cruise engine on-line;
- cruise engine clutch engaged; and
- shaft not locked.

The drive mode schedule indicator provides information to the operator and to the PSC as to which combination of schedules (propeller pitch, shaft revolutions per minute, and power) are to be utilized. For a COGAG machinery arrangement, there are a number of

Table 9.1 Gas turbine propulsion system states.

State	Ready to start	Running	Connected to controller
Standby	Yes	No	No
On	N/A	Yes	No
On-line	N/A	Yes	Yes
Off	No	No	No
Normal stop	No	Yes/no	No
Emergency stop	No	No	No

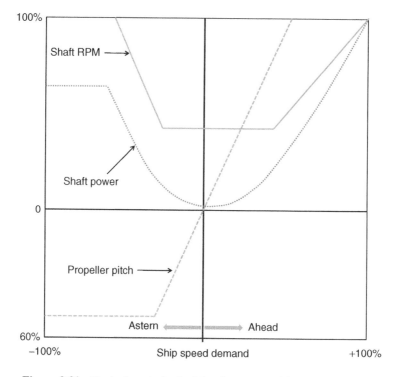

Figure 9.21 Typical control schedules for a gas-turbine-powered ship.

combinations of machinery that are possible (and permissible). Each of these combinations of machinery will have a set of operating schedules calculated to obtain the best performance from the machinery. A typical set of schedules for a particular drive mode is shown in Figure 9.21.

The limitation of propeller pitch and shaft power that is associated with astern maneuvers can be noted from Figure 9.21. These limits are normally associated with a limitation of propeller shaft and blade spindle torques due to the messy flow conditions associated with reverse pitch. This is especially so during a 'crash astern' maneuver.

9.6.4 Propulsion System Controller

For the purposes of this discussion, the propulsion control consists of propeller pitch control and speed control of each gas generator component of the gas turbine. It should be self evident that multiple engines driving will result in a control loop for each engine; however, while each may be trimmed somewhat differently, the general form will be the same.

The propeller pitch control is usually a simple forward (open) loop control which establishes a propeller pitch demand in accordance with the previously determined schedule (e.g., Figure 9.21). This demand signal is fed to the propeller pitch actuator which is

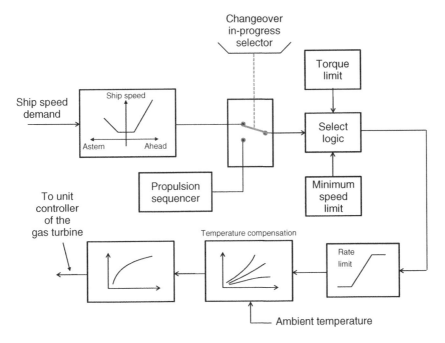

Figure 9.22   Typical gas turbine propulsion control logic.

normally supplied by the manufacturer of this component. Actual pitch adjustments are achieved through a servo system that provides hydraulic oil pressure to a central piston which, in turn, is mechanically linked to each propeller blade spindle.

The control of power from each gas turbine is achieved by the conversion of the ship's speed demand to a shaft power demand in accordance with the appropriate schedule as shown in Figure 9.21. This shaft power demand is then converted through a series of control blocks to a gas generator speed demand which is then presented to the gas turbine unit controller. As discussed previously, this controller is usually supplied by the gas turbine manufacturer which provides detailed logic governing fuel schedules and variable geometry schedules as required for the gas turbine supplied.

In general terms, these control functions will follow the descriptions presented in Chapter 3 of this book. A block diagram of a typical gas turbine propulsion control loop is shown in Figure 9.22.

As noted in the figure, there are several limiters and compensators that apply to the overall loop. Of these, shaft torque is probably the most significant. Downstream of the gas turbine is a large gearbox which must transform the torque/speed from the gas turbine to torque/speed appropriate to a propeller turning at roughly 1/20 of the gas turbine output. There have been some notable catastrophes with marine gearboxes that have alerted the engineering community to the need for caution in dealing with these propulsion components.

9.6.5 Propulsion System Sequencer

The PSS is that part of the overall propulsion control of a ship which executes the drive mode changeovers. These changeovers include putting any of the gas turbines into and out of drive mode and managing the transition from one drive mode to another. This includes starting and/or stopping any of the engines. During the cooling-down period of an engine which has been disengaged, the sequencer remains in control of that engine. During the transition from one drive mode to another, the sequencer will assume control of the ship's speed demand signal and will adjust shaft speeds and load applications so that a 'bumpless' transfer from one drive mode to another is executed. Once the transfer is complete, the PSC resumes command of the ship speed demand input.

For the example of the COGAG propulsion configuration considered thus far, the allowable drive modes are as follows:

- shaft trailing;
- boost engine driving at idle;
- boost engine driving;
- cruise engine driving at idle;
- cruise engine driving; and
- cruise and boost engines driving.

The move from one drive mode to another requires a series of commands to be executed by the sequencer. For example, to change the drive mode from 'shaft trailing' to 'boost engine driving' would require the following sequential commands:

- boost engine on; and
- boost engine on-line.

To switch to the cruise engine driving will require the following:

- cruise engine on;
- boost engine normal stop; and
- cruise engine on-line.

These examples indicate the general format of sequencer logic. Each command will activate unit controllers during start-up and shutdown modes such that the required warm-up and cool-down periods are honored. They also will activate, through the start sequence, the lube oil system and fuel supply system, enclosure ventilation system, and all other ancillaries. However, in modern applications of marine gas turbines these are handled entirely by the gas turbine unit controller.

9.7 Concluding Commentary

The material presented in this chapter illustrates the fundamental differences in machinery complexity and operational philosophies between the marine gas turbine propulsion systems associated with the typical naval surface ship and the corresponding aircraft propulsion systems. Factors which drive these differences include:

- the sea-going environment;
- the requirement for long-term autonomous operation at sea;
- the complex machinery involving multiple engines and multiple control modes; and
- the critical dependence upon the propulsion system to deliver the power required quickly and reliably in support of overall ship performance in hostile situations.

Even with the established capabilities of modern control technologies, there remains a strong need to maintain a high degree of manual control and oversight. However, as system complexities have grown, much of the system and machinery status monitoring and reporting has been delegated to automation in the interests of reduced crew workload.

References

1. Williams, D.E. (1995) The Naval Gas Turbine. Part 1, Maritime Defence, January 1995.
2. Carleton, R.S. and Weinert, E.P. (1989) Historical Review of the Development and Use of Marine Gas Turbines by the US Navy, ASME 89-GT-23D.
3. Shepherd, D.G. (1956) *Principles of Turbomachinery*, MacMillan Company, New York.
4. Brady, C.O. (1987) The general electric LM 2500 marine gas turbine engine. Presentation to St. Lawrence College, Kingston Ontario, January 1987.
5. Williams, D.E. (1981) Progress in the Development of the Marine Spey SM1A. ASME 81-GT-186.
6. Cowley, J. (1992) *The Running and Maintenance of Marine Machinery*, Institution of Marine Engineers.

10

Prognostics and Health Monitoring Systems

Prognostics and Health Monitoring (PHM) of modern propulsion systems has emerged as an important area of technological development in recent years. Some observations of the changes that have occurred that are driving this development offer important considerations that are well worth keeping in mind from a propulsion systems perspective.

It is somewhat arbitrary and largely a matter of opinion when the changes began; however, it was about 1970 that the 'big' jet era began. This period saw the introduction of wide body passenger airplanes of which the Boeing 747 was the first. This was powered by the Pratt & Whitney JT9D, a new high-bypass-ratio dual-stream turbofan engine. General Electric had already fielded the TF-39 engine on the US military transport, in the form of the Lockheed C5-A. In 1971, Rolls-Royce was forced into receivership because of extraordinary costs incurred in the development of the RB-211 high-bypass turbofan engines for the Lockheed L1011 commercial jetliner.

The 1970s also ushered in the Arab Oil Embargo where the Organization of the Petroleum Exporting Countries (OPEC) cartel controlled the production of crude oil which resulted in rapid and unpredictable changes in the world price for crude oil. Starting from a price of about $4.00 per barrel, the price of oil at the time of writing is in the region of $70–90 US dollars per barrel and is predicted to continue to rise on the basis of increasing worldwide demand and uncertainty in sources of supply. Using the following very crude estimates for a modern twin-engine airplane, we can make a useful estimate of the cost of fuel for operations:

- 3000 flying hours per year,
- average fuel consumption of 10 000 lb/h (5000 lb/h per engine),
- average fuel cost of $1.20 per l ($5.50/gal).

The above assumptions result in an annual cost in fuel alone of about $14 million per airplane per year ($7 million per engine per year). In other words, an engine consumes

Gas Turbine Propulsion Systems, First Edition. Bernie MacIsaac and Roy Langton.
© 2011 John Wiley & Sons, Ltd. Published 2011 by John Wiley & Sons, Ltd.

close to its capital value per year in fuel. It is also interesting to note that a pipeline stationary gas turbine engine, which is typically an aero-derivative gas turbine, consumes about twice its capital value in fuel per year and that a 3% drop in efficiency is generally regarded as enough to warrant an overhaul.

Considering the same period of time (1970–2010), it is impossible to be unaware of the rise of the environmental movement and the public concern over the contribution of greenhouse gases (CO_2 and others) to the problems of global warming and climate change. Nobody can argue with the fact that CO_2 is produced by a jet engine. Similarly, no one can realistically argue against the continuance of the airline industry as the principal mode of international travel. Until a completely different mode of propulsion is discovered, the jet engine will remain the workhorse of the airline industry and development will continue to improve its overall efficiency. Similarly, changes in efficiency over time while on-wing are an important factor both in keeping operating costs under control and minimizing its contribution to greenhouse gases.

In addition to the overall improvements made in the jet engine over the past 40 years related to performance and reliability, the emergence of the microprocessor as the technology of choice for controls is undeniable. While this new technology has not resulted in any significant changes in the fundamental principles of gas turbine control, its inherent flexibility has resulted in a substantial increase in functionality. This has resulted in reduced pilot workload and provides a means of collecting data to facilitate continuous improvements to be wrought in all supporting operations of the aircraft. Nowhere is this more evident than in the management of the operations and maintenance tasks associated with flight safety.

Finally, the period 1970–2010 has resulted in changes in the cost of living of the order of 6.5–7.0. To the extent that wages have kept pace with this change, we are forced to recognize that labor is an increasingly large component of most operations. Competitive forces therefore drive us to seek improvements in every aspect of flight operations which will reduce the costs of both material and labor. Stated simply, we want to extract as much value as possible from resources in the interest of economic advantages in the marketplaces that the jet engine serves.

Using the simple adage that knowledge is power, the concepts motivating the development of PHM for propulsion systems are as follows.

1. It is necessary to obtain data to inform us of the current status of each machine.
2. There are cost areas in operations and maintenance that have prospects for significant improvements if additional information could be made available.
3. The timelines of the data collection and conversion to useful information is critical to the success of any such effort.
4. Such developments have important elements of feedback to the designers. Such information provides a basis for improvement in the future designs (lessons learned).

The following sections will attempt to describe the elements of a condition-based maintenance concept and how it can be used to bring about improvements in such an important industry.

10.1   Basic Concepts in Engine Operational Support Systems

Before a discussion of the specifics of an engine PHM system can begin, the underlying concepts of engine operational support must be understood. As stated in Chapter 2, the two pacing technologies that have continued to drive and/or limit progress in jet engine design are materials and performance. These limitations manifest themselves in flight operations and might be described as problems that must be dealt with throughout the life of the machine. These will be discussed in the following sections.

10.1.1   Material Life Limits

It must be recognized that the materials used to make the rotating components of a gas turbine are typically operating close to their thermal stress limits. For example, the radial stress experienced by a compressor or turbine disc passes through its elastic limit during every acceleration to full thrust. Similarly, blades are subjected to a combination of stress and temperature such that the material is at or near the limits of its capability in stress and creep.

The philosophy supporting such a design concept is that the weight of the engine must be kept to an absolute minimum. The cost of unnecessary weight results in a significant loss of aircraft performance together with the additional cost of fuel required to be carried in support of the mission. Using this as a basic reality, it is fair to say that a gas turbine is designed to operate at or close to the material limits of its major components and anything more is judged to be too heavy.

This philosophy drives the designers to minimize weight and then to test the design to determine the acceptable life that can be extracted from the component before it must be retired. These tests, generally referred to as accelerated mission tests, are conducted in spin pits where the major rotating components are repeatedly started, run up to speed, and stopped until failure occurs. Sufficient components are sacrificed to this process to obtain a statistically acceptable estimate of the part life in service.

The consequences of a disc failure in flight can amount to a catastrophe. It is possible to design shrouds which will contain the loss of a fan or turbine blade; however, the energy associated with a disc failure is so high that no practical means of containing it with certainty is known today. Figures 10.1 and 10.2 provide graphic evidence of the consequences of a disc failure in flight [1].

The design and operational philosophy is therefore to never allow a disc failure in flight. To achieve this goal, the accelerated mission tests from which the life estimates are made are discounted by a factor selected to ensure that such an event does not occur. For the cold end components (compressors and fans), this discount is of the order of 3. For the hot end components it is of the order of 5. These write-downs of life expectancy reflect the following uncertainties:

1. the inevitable uncertainty of the presence of an imperfection in the forging from which the part is made at which a crack can be initiated; and
2. the uncertainty that the accelerated mission test accurately represents the real environment of engine operation.

Figure 10.1   Nacelle of a DC-9 following a JT-8D turbine disc failure (courtesy of the NTSB).

Figure 10.2   Part of engine rotor of the failed JT-8D (courtesy of the NTSB).

Table 10.1 Typical life limits for aircraft turbine engine components.

Section	Article	Typical life (hr)
Fan	Fan discs	5000
	Fan blades	1000
Compressor	Compressor discs	5000
	Compressor blades	6500
Compressor	Combustor casing	4500
	Combustor fuel nozzles	3000
HP turbine	Turbine nozzle vanes	–
	Turbine blades	1300
	Turbine discs	3500
LP turbine	Turbine blades	4500
	Turbine discs	4000

A similar set of arguments applies to the stationary components of the combustion chamber, turbines, and exhaust sections. These components are subject to thermal creep, oxidation, and thermal stress fatigue. A combination of failure modes is therefore ever present and these parts must be replaced before any internal failure can occur. The reader is left to contemplate the fate of a downstream turbine blade row in the event of a fatigue failure of any upstream component.

A summary observation of the above description is that there are a number of components of a jet engine which have parts that are subject to replacement after some measure of usage. A representative set of lifed parts for a modern jet engine is provided in Table 10.1. These are but representative and another engine may have additional lifed components; the specified times will almost certainly be different.

In order for an engine to obtain a flight certification, the manufacturer must present data describing the anticipated life of each component and the justification for the life limits imposed. Once certification is obtained, any operation of the engine becomes subject to these rules and any operation outside of them is against the law.

A glance at the list of parts lives shown in Table 10.1 indicates a dilemma for any operator. Life consumption is tied to operation, during which the whole engine is put to work. The law demands that the engine be removed from service at the expiry of the life of any one of the parts listed. Taken to its extreme, it is obvious that the engine will have to be torn down rather frequently if all of the life of all of the components is to be taken to their respective limits. While parts are very expensive, the continuous removal of engines for parts replacement is also a major financial burden. Such a state of affairs has produced several operating practices that are of importance to this discussion.

First, it is obvious that the number of shop visits must be managed as a function of time. After all, there are only so many positions in any facility that can accommodate an engine. This suggests the obvious: establishment of a utilization profile for each airplane/engine such that removals are occurring at a rate that can be accommodated by the shop.

Secondly, a tradeoff must be established between the value of the remaining life of a component and the cost of a shop visit/engine teardown/part replacement. This suggests the establishment of a 'window of opportunity' in time or life consumption units which

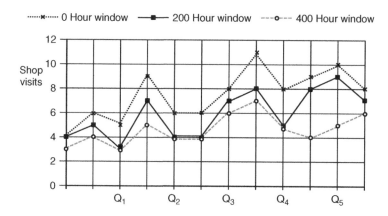

Figure 10.3   Shop visit rate for three opportunity windows.

is used to determine which parts will be replaced at a specific shop visit. At the point where a specific part has reached its limit and therefore must be removed, the operator will observe other parts that are within the window; that is, they are close enough to their limit to warrant replacement at the same time as the part that triggered the visit by the engine to the repair shop. Figure 10.3 shows the consequences of a number of 'windows' of remaining parts lives on the shop visit rate of a typical fleet of aircraft.

As can be seen from the figure, at the expense of wasting some of the remaining life of those parts which are near but not yet at their life expiry, the problem of shop visit rate can be kept under control. Note that failure to keep shop visit rate under control drives an operational requirement for more spare engines and an extended pipeline of 'engines or components in process', both of which represent very large investments in spares.

10.1.2 Performance-related Issues

The modern civil aircraft engine is the subject of substantial design and development effort to obtain continuous improvements in fuel efficiency. Fuel burn is of paramount importance to an airline. While cost and environmental concerns continue to be drivers in this endeavor, it is equally important that the engine be able to obtain high efficiency throughout its life on the airplane. Here, the certificating authority is silent. After all, such a body is set up to protect the public safety. So long as this is achieved, the economic performance of an airplane and its propulsion system is entirely the concern of the operation. In this regard, there are a number of engine degradation modes which do not directly or immediately affect safety but which must inevitably affect fuel consumption.

For example, the nozzles of the turbines are subject to thermal creep and to changes in blade profile as a consequence of operational cycles. These effects result in geometric changes in the nozzle blade profile which has a direct effect on turbine efficiency. Figures 10.4 and 10.5 depict the results of tests undertaken to assess the consequences of a number of common operational degradation modes on turbine efficiency and flow capacity. These tests clearly indicate the changes that can and do occur in high-performance

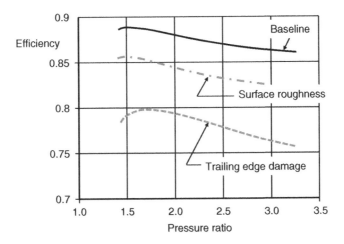

Figure 10.4   Effect of damage on turbine efficiency.

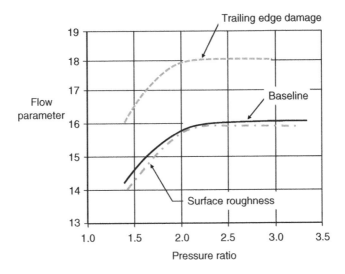

Figure 10.5   Effect of damage of turbine flow capacity.

engines as a consequence of thermally induced damage to turbine blading (both nozzles and rotating blades).

More subtle forms of performance changes are also present in the operation of a high-performance engine. Seal leakage robs the engine of efficiency in both compressor and turbine sections. These leakages can represent several points in overall thermal efficiency of the engine.

If the engine is equipped with variable stators on any number of compressor stages, the actuators for these vanes are subject to wear and friction. This can result in loss of performance of the compressor and, in a worst-case scenario, compressor surge.

Compressor blade erosion and/or fouling is a problem on certain types of engine operation. The gas turbine is, after all, an air-breathing engine. While it is presumed that the bulk of its operation is at altitudes where air pollution is insignificant, the airplane most probably lands and takes off in an environment less than pristine. Blade erosion and foreign object damage can occur on a continuous basis. Most of this damage is slight but progressive, and takes a toll on overall engine performance without necessarily putting the safety of the engine in question.

10.1.3 Unscheduled Events

Regardless of the diligence of operators and maintainers, there will be situations which result in unscheduled shutdowns of the engine. These may be substantial events such as the sudden failure of a bearing or an in-flight flame-out leading to a loss of thrust. They may be little more than a nuisance alarm as a warning, which causes a delay upon departure. Nevertheless, there is a statistical likelihood that unexpected events will result in loss of operation and maintenance work which must be borne directly or indirectly by the operator.

The foregoing provides some insight into the world of jet engine operations. As can be anticipated, this environment requires actions to taken by operations and maintenance personnel at several levels as follows:

- flight line operations (replace and adjust);
- local repair facilities (test, repair some components); and
- remotely located (deep strip refurbishment).

Some of the work required to support a fleet of aircraft is driven by the scheduled removal of lifed parts. Other work is a direct consequence of either performance-related degradations or response to operational situations. The basic notion of PHM is the systematic collection of relevant data which can result in improved engine performance, better utilization of the engine and its components, and overall reduction in operational costs.

10.2 The Role of Design in Engine Maintenance

Much of the practice of engine maintenance is specified by the engine designer through his choice of materials and cycle parameters which define temperatures, pressures, and rotational speeds of shafts and thus stress conditions of individual engine components. Throughout the design process, the design engineer is cognizant of the consequences of his choices regarding operations and maintenance. The so-called 'ilities' (reliability, maintainability, and availability) are, in truth, demands made by the users of these machines which drive the competitive process during the selection of an engine for a new application. In turn, the designer strives to meet these demands subject to a maintenance regimen that he defines.

10.2.1 Reliability

Much has been written on the subject of reliability [2–4]. Stated simply, it is a statistically derived term which expresses the likelihood that a part will survive for a given period of time subject to a known set of external conditions. This is expressed mathematically using a probability density function, as shown in Figure 10.6.

Figure 10.6 implies a large number of tests; a histogram was created which ranked the data according to a 'frequency of occurrence' f, usually expressed as a percentage of the population. The total area under the curve will therefore always be 1 or 100 depending on the way the percentage is expressed. However, the area under any part of the curve is defined by integrating the probability density function with respect to time up to some specified value of time t. We therefore have

$$F(t) = \int_0^t f(t)\,dt \tag{10.1}$$

Since reliability R essentially expresses the likelihood of survival, it is that part of the total area not included in $F(t)$. In other words,

$$R(t) = 1 - F(t) \tag{10.2}$$

Examining Figure 10.6, and setting a life expectancy of t_1, we see that $F(t_1)$ is a relatively small portion of the total area under the curve and that $R(t_1)$ represents a highly reliable part. However, if we set a life expectancy of t_2 we will be disappointed since $F(t_2)$ captures a very large portion of the failures and $R(t_2)$ will be quite small.

In developing the data upon which the reliability of a part is based, we cannot know *a priori* what the shape of the probability density function will look like. However, extensive testing over many years provides the guidance summarized by Table 10.2 below.

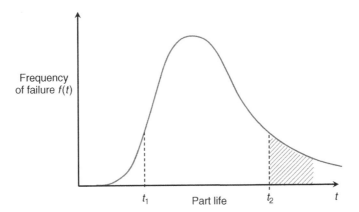

Figure 10.6 Probability density function.

Table 10.2 Distribution and application.

Probability density function	Application
Normal distribution	Material properties:
	Hardness
	Electrical capacitance
	Tensile strength, etc.
Weibull distribution	Life expectancy:
	Early
	Random
	Late
Exponential distribution	Life of assemblies
Binomial distribution	Appearance of defects

Such guidance is important to the engine developer because he must base his life predictions on a finite number of tests. Knowing the general shape of the statistical phenomenon allows an estimate to be made with much fewer tests. For example, the Weibull distribution is used for all parts subject to low cycle fatigue (LCF) failures. The complete definition of this density function is

$$f(t) = \left[\frac{b}{\theta - t_o} \left(\frac{t - t_o}{\theta - t_o} \right)^{b-1} \right] \exp \left[- \left(\frac{t - t_o}{\theta - t_o} \right)^{b} \right] \tag{10.3}$$

where t_0 is the expected minimum value of t, b is shape parameter (Weibull slope) and θ is a characteristic value (scale parameter).

In general terms, this is an extremely useful probability function that can be used to represent many situations. It is observed that the shape parameter b obtained by plotting actual test data displays different portions of the life of a part as follows:

- $b < 1$: infant mortality;
- $b = 1$: random failures; and
- $b > 1$: old age.

A typical Weibull plot is shown in Figure 10.7. Once the shape of the distribution is known, the calculation or prediction of reliability for a given life is a straightforward application of the mathematical formulation. As mentioned previously, the stated life for flight safety is a discounted value of the life obtained from the tests.

The application of reliability statistics to assemblies of parts dictates that the overall reliability is the product of the reliability of the individual parts as follows:

$$R_s(t) = R_1(t) \times R_2(t) \times R_3(t) \times \ldots \ldots R_n(t) \tag{10.4}$$

where n is the number of parts involved. Note that this is the common situation for any machine where the failure of any one part will cause the machine to stop. This is the principal argument for redundancy in flight critical hardware.

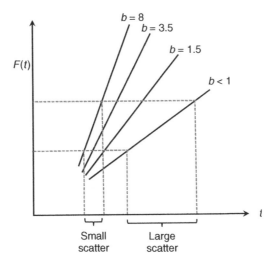

Figure 10.7   Weibull distributions.

10.2.2 *Maintainability*

In general terms, maintainability is a measure of the ease or difficulty of keeping a piece of equipment in an operational state. More formally, maintainability has the definition:

> The ability of an item to be retained in or restored to specific conditions when maintenance is performed by personnel having specified skill levels, using prescribed procedures and resources at each prescribed level of maintenance and repair.

Ultimately, maintainability is measured in monetary terms; however, it is usually described as the 'time' required to go through the above-mentioned procedures and the number of man-hours of effort to do so.

To design a machine for maintainability is to study and assess the time required to maintain the machine through all possible failure modes and, from that assessment, attempt to arrange the design to minimize the cost of each maintenance action. Application of this concept to the jet engine requires awareness of the operational theater and the organization of maintenance in the theater.

Regardless of the nature of the failure, all maintenance actions can be broken down into individual tasks. These tasks must be timed in accordance with human effort and available equipment. The usual method of presenting the work of maintenance is as shown in Figure 10.8.

It is common practice, as noted in the figure, to account for the item repair separately from the so-called system work. However, examination of the diagram yields an awareness that any malfunction will result in stoppage of the machine and the mobilization of a crew under the assumption that a specific malfunction has occurred. As noted, there is a preparation time where the crew will be assembling required material to address the assumed problem.

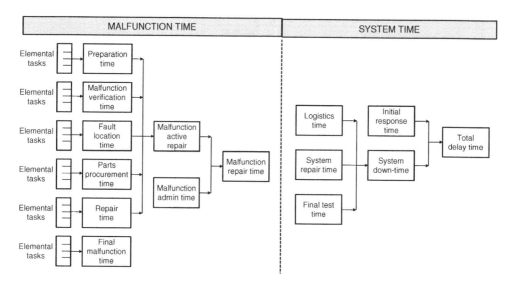

Figure 10.8 Structure of time elements in a 'fix'.

Once ready, the first task is the verification of the malfunction. This involves whatever repeat tests are required to verify that the malfunction is as reported. There are usually a number of possible candidates, all of which could cause the malfunction. The crew is expected to undertake all tests required to isolate the fault to a specific part or assembly.

Once the fault has been isolated to a particular component, the magnitude of the repair task is known and the parts required to complete the repair can be procured and made ready for installation. This allows the work to move on to the actual repair of the component and reassembly of the complete machine.

Finally, the crew must adjust, calibrate, and perform all tests necessary to ensure that the repair has been successful.

Note that Figure 10.8 indicates that time and motion studies down to what is referred to as elemental tasks are required to build up the total time. These elemental tasks include such details as the removal of a bolt and often form the basis of design ideas to improve the maintainability of the machine. This was very noticeable in early jet engines where multiple pipes and tubes had to be removed simply to get at an adjustment or inspection port. The introduction of modular designs forced a considerable simplification of external piping on most machines.

Referring again to Figure 10.8, note that after the component repair is completed there remain a number of so-called 'system times' that must be accounted for. These include records of maintenance action, delays in initial response, and final system-level tests, all of which contribute to the overall delay time. From this and the concepts of the previous section, we can define three commonly used terms as follows:

Mean time between failures (MTBFs). This term is directly related to the component and system reliability parameters and is related to the design choices made for the machine assuming full knowledge of the operating environment.

Mean time to repair (MTTR). This term is obtained from the maintainability analysis described above. It usually implies only the work required to effect the repair under optimum conditions.

Mean delay time (MDT). This term is also obtained from the maintainability analysis but includes realistic times for logistics. Thus, while MDT could be identical to MTTR, it is usually longer and includes times that are associated with the organization of maintenance at the operator location and are thus either directly or indirectly under his control.

10.2.3 Availability

The availability of a machine AV is defined as the amount of time that a machine is available for operations as a fraction of the total time. This is defined mathematically as:

$$AV = \frac{MTBF}{MTBF + MDT} \tag{10.5}$$

Obviously all time between failures that is not consumed by the repair process is available for operations. Thus, the overall reliability of the machine determines the MTBF (any part) and is principally a design issue over which the operator may have some control depending on the duty of the engine. The MDT contains work times that may yield to clever operator actions; however, it is often dominated by logistics over which he has considerable control.

We can rewrite the expression for availability in the form:

$$MDT = \left\{ \frac{1 - AV}{AV} \right\} MTBF. \tag{10.6}$$

We now have a simple linear relationship with a fixed slope for any given value of AV, which is shown in graphical form in Figure 10.9. The interpretation of this plot is quite interesting since it shows quite clearly the relationship between the quality of the design and the logistics of operational support.

On the abscissa, the independent variable is MTBF. As was discussed previously, this is a statistical function with a probability density function centered on the mean. MTBF is directly related to individual component reliability, with the overall system reliability determined by the product of the individual component reliabilities. Within the context of a given operational duty, the MTBF of the system is directly related to the design and to the compromises the designer has made between weight and stress. It is also worth noting that, for a set of individual components whose MTBF are roughly the same, the specific reason for system failure is distributed roughly equally between all such components. This statement has important consequences for the determination of maintainability.

The ordinate in the plot shown in Figure 10.9 is the MDT. This parameter includes all times that may be allocated to the individual activities required to put the machine back into service. In broad terms, the following tasks may be seen to apply:

- observe and report malfunction;
- tests to confirm the malfunction;

- disassembly of unit;
- repair/replacement of component;
- reassembly of unit;
- set-up and adjustment of system; and
- tests to confirm successful repair.

As can be seen from the above, the nature of the malfunction (component, position in the engine, etc.) has a profound effort on the overall time required to complete a disassembly/repair/reassembly sequence. For example, the repair/replacement of an externally mounted part (e.g., inlet temperature sensor) bears little relationship to the replacement of a main bearing. It may therefore be seen that the move to modular designs which occurred in the 1970/1980s time frame was a direct response to improving availability by allowing a more rapid replacement of a module rather than a full engine teardown and rebuild. This is an example of how the design can affect the MDT for a given repair.

The MDT includes times associated with overhaul and maintenance (O&M) issues that are not directly under the control of a designer. Again, an example is instructive. In the event of an engine over-temperature on a modern fighter engine, it is commonplace for the ground crew to suspect the engine controller. Because this is an externally mounted item (a so-called line replaceable unit or LRU) it is a favorite suspect resulting in its removal/replacement simply as a diagnostic trial. In the event that the cause of the problem is elsewhere (e.g., temperature sensor), considerable time and expense will have been consumed by such a wrong diagnosis.

The above is simply an example of the operator's potential to contribute unprofitably to the MDT. Others include availability of qualified technicians, availability and location of spares, availability and location of tools, and so on. In a well-organized effort, the total delay time can approach the time to repair or the 'malfunction time' specified in Figure 10.8. In a poorly organized effort, the total delay time can be much longer.

Returning to Figure 10.9, the greater the MDT, the poorer is the availability. As noted in the figure, for an arbitrary MTBF of 1000 hours the MDT cannot exceed 111.1 hours in

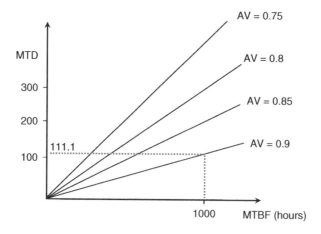

Figure 10.9 The availability function.

order to achieve a system availability of 0.9 or better. To an engine designer, improving systems availability means improving individual component reliability and/or producing an engine design that is easily diagnosed and repaired. To an engine operator, improving system availability means more highly trained technicians and improved logistics.

10.2.4 *Failure Mode, Effects, and Criticality Analysis*

Failure mode, effects, and criticality analysis (FMECA) forms the basis of the designer specification of maintenance regimens. From the results of this analysis, together with his knowledge of the design and the anticipated operational environment, he is expected to produce specific maintenance instructions as well as a definition of what is commonly referred to as 'operational procedures and permissives'. This is shown diagrammatically in Figure 10.10.

As the diagram indicates, the results of the engine design effort is the specification of a bill of materials and, as a consequence of LCF and related factors governing the life of parts, an estimate of mean time to failure of all parts is made.

A failure mode analysis is a systematic inquiry into all of the possible failure modes of each and every part on the bill of materials. The principal questions that must be posed in such an analysis are as follows.

1. How can this part fail?
2. What are the consequences of such a failure?
3. What action is to be taken to prevent such a failure?

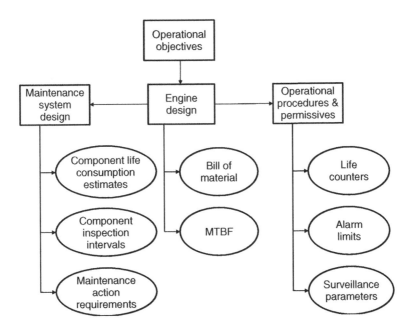

Figure 10.10   Engine design outputs.

The answers to these questions drive the specification of operational procedures and permissives shown in Figure 10.10. By combining the results of the failure modes and effects analysis with the anticipated mean time to failure of each part, the maintenance system design can be undertaken resulting in a maintenance design document. As noted in Figure 10.10, the maintenance system design must estimate the rate of consumption of component life. From these estimates, the designer will set inspection intervals, parts removal intervals, and all maintenance-related actions that directly affect issues of flight safety.

The estimate of component life consumption is a very inexact science. Prior to the 1970s, engine manufacturers specified the life of components in terms of flight hours. Operators maintained logs of flight hours and engines were removed from service when a specific component reached its allowable hours of usage. In many cases, this was simplified to an overall life expectancy for the engine at which point it was sent for overhaul as required.

As more experience was gained with specific engines, it became apparent that simple operating hours was a very crude measure of life consumption. For example, the number of times an engine is started clearly affects its life in a negative sense. Thus short-haul aircraft were observed to have a harder duty than a long-haul airplane.

A similar observation could be made for those operations that demanded frequent changes in power level. For example, helicopter operation with frequent takeoffs, landings, and maneuvers involving pick-up and delivery of goods and personnel do not involve engine starts/stops, but most certainly involve many changes in power level.

These experiences led to the modification of the flight hours to account for a mission severity factor. This in turn led to recognition by operators that if they could devise operational regimes that treated the engine more gently, it could improve its life expectancy; however, this only benefited the operator if he could get some relief on the lives of components through the flight regulators.

As time went on and the power of computers was brought to bear on the issue of estimating life consumption, the designers began to specify component lives using algorithms which recognized more detailed definitions of the actual operating environments. Today, many of the major components of an engine have lives which are specified in terms of a number of engine measurements. For example, a turbine disc life might be specified in terms of:

- time at different speeds;
- a cycle temperature such as turbine exit temperature; or
- a cycle pressure such as compressor delivery pressure.

By developing a 'life consumption' parameter which is functionally linked to key engine cycle parameters, the rate of life consumption can be measured much more accurately. This has many benefits: the designer can optimize his design confident that specific parameters will continue to flow to his desk as engines enter into service, and the operator can more accurately track the usage of his equipment leading to streamlined operations and maintenance practices.

These developments have become possible and indeed routine by the introduction of microprocessor-based recorders that are fitted to modern engines for purposes of collecting critical data on engine operations. Initially, these 'black boxes' were only considered to be intended for tracking the life consumption indices associated with LCF of specific lifed components. However, the benefits of data which provided records of flight events which in turn led to earlier and more accurate diagnosis of system problems was soon made apparent. These capabilities have led to concepts which are commonly referred to as:

- retirement for cause;
- condition-based operations; or
- condition-based maintenance.

All have a common basis for their development: the existence and/or availability of data which can be used to make informed decisions about the operations and maintenance of systems. This development is referred to as prognostics and health monitoring and is now an active component of any engine development program. The elements of PHM will be discussed in the following sections.

10.3   Prognostics and Health Monitoring (PHM)

While the technological basis for PHM is the availability of low-cost electronics, the rationale for its development is very much an economic one. Fuel is expensive, parts are expensive, people are expensive, and competition is fierce. In the text opening this chapter, annual fuel costs for a modern jet engine were estimated to be $7 million. A 3% increase in consumption translates to $210 000 per annum. Spread over an entire fleet, such costs strongly suggest that the monitoring of performance is worthwhile. Similarly, a set of turbine blades for a single-turbine stage costs around $500 000. If the consequence of higher fuel consumption is an operating temperature increase of the order 10 °C, the life of the turbine blade set will be consumed in roughly half the time.

Figure 10.11 is a depiction of a typical aircraft support organization. The figure indicates three recognizable organizations, each operating at a physical location that is convenient to the execution of the work involved.

The flight line maintenance group is primarily concerned with the following tasks

- servicing the aircraft refueling process;
- inspections of physical states including common fault indicators; and
- replacement of commonly replaceable components such as sensors, actuators, indicator lights, and so on.

It is at the flight line that technicians engage in the diagnosis of reported faults and have limited time available to do so.

The main operating base is typically owned by the operator and it is to this location that a grounded aircraft (or an engine) is brought for more extensive work. It is at this location that the following tasks are executed:

- major inspections;
- major component replacement;
- some component repairs; or
- data management for the fleet.

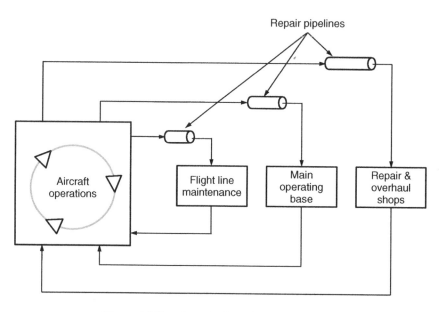

Figure 10.11   Organization of aircraft support.

Finally, the Repair & Overhaul (R&O) shops may be geographically distributed and may or may not be owned and controlled by the operator. It is to these shops that engines and/or component modules go for complete refurbishment.

The introduction of additional measurements of performance and mechanical health afford the operator an opportunity to materially affect the costs of operations that are described in Figure 10.11. Indeed, it is believed that the work at the flight line can be reduced to simple servicing of the aircraft and that all problems that can be rectified with simple parts replacement can be done opportunistically as the aircraft 'passes through' its main operating base; a concept which has yet to be fully demonstrated.

10.3.1   The Concept of a Diagnostic Algorithm

The underlying assumption of PHM is that fitment of specific sensors to the engine will lead to an ability to isolate a specific problem as it emerges, in sufficient time to allow the aircraft to be serviced/repaired quickly and efficiently without a major disruption to the operations.

Perhaps it would be more accurate to state that the aircraft or other vehicle is the productive unit which must be kept in service; however, the engine is a major contributor to loss of operation so this discussion will focus on that component. Regardless of the application, the key technological development associated with PHM is the capability to interpret measurements which ensure accuracy, lack of ambiguity, and can be handled by an onboard computer. This is shown notionally in Figure 10.12.

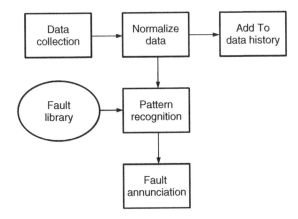

Figure 10.12 Basic notion of fault diagnosis.

Without exception, all data that form inputs to an on-board diagnosis must be interpreted in several ways as follows.

1. Do the data indicate a real problem?
2. Does the severity of the problem warrant action on the part of the operator?
3. How much operational time is left before a shutdown is mandatory?

For example, if the information is advisory or (worse yet) doubtful and thus requires a crew of technicians to perform additional tests to confirm the fault, then much of the advantage of PHM is lost. The notion of a Fault Library (as depicted in Figure 10.12) is therefore fundamental to the notion of PHM.

10.3.2 Qualification of a Fault Indicator

The starting point in the development of a PHM program is the selection of and qualification of a fault indicator. The reader is reminded that a gas turbine comprises a number of conjoined thermodynamic processes, each of which require some half-dozen variables to define the operating point. For example, a compression process requires an accurate definition of inlet conditions (pressure, temperature flow rate, and rotational speed) as well as the corresponding outlet conditions (pressure, temperature, and flow rate) to fully define its operation. It becomes more complicated if there are bleed takeoffs.

Bearing in mind that the compression process itself is achieved by a mechanical arrangement of multiple blade rows, each of which is shaped to contribute to the overall compression achieved by the machine, then we obtain some sense of the complexity of a diagnostic process. In the limit, we would like to know precisely what changes have occurred to what blade row to have caused a worrisome deviation in overall performance. If we now add the other thermodynamic processes to obtain a complete working

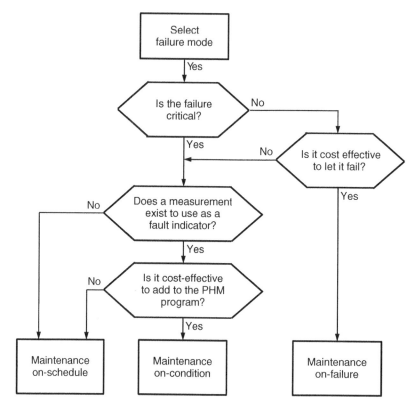

Figure 10.13 Failure mode analysis applied to operation and maintenance.

engine and realize that these components are aerodynamically, thermodynamically, and mechanically linked, the true dimensionality of the problem becomes clear. It is therefore imperative to select and qualify useful fault indicators that can provide information about the condition of the machine to the operation and maintenance personnel.

The development of a qualified condition indicator requires a return to the FMECA as indicated in Figure 10.13. As can be seen from the figure, there are essentially two underlying objectives to the analysis: safety of operations and cost of operations. All failure modes must be considered; however, each failure mode must be judged in terms of safety and cost.

As indicated, the result of such an analysis is the emergence of a number of failure modes which do not affect safety and for which a systematic tracking of some indicator is more cost effective than adherence to a fixed schedule. The diagram also suggests that there is a continuous effort to obtain improvements. Indeed, the current thrust of the so-called 'lean initiative' dictates continuous improvement. Not only must there be an effort to improve the schedules for maintenance, but there must also be a systematic effort to develop and qualify fault indicators that can be made useful in the management of the engine over its service life.

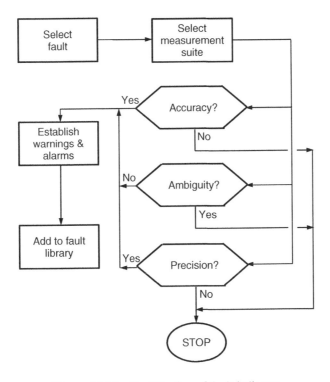

Figure 10.14 Qualification of fault indicator.

An overall diagram of tasks involved in the qualification of a fault indicator is shown in Figure 10.14. The process begins with the selection of the fault in question. This implies that either:

1. the fault is serious; or
2. the fault is economically worth the effort to develop an indicator for it.

The selection of the method of measurement is a non-trivial exercise involving both the underlying physical phenomenon and the practicality of its inclusion in what must be made an automatic procedure.

Following selection of the measurement suite, the accuracy of the measurement must be sufficient to expose the existence of the fault. This question has elements of both safety and economics attached to it. It also challenges the developer to obtain either test data and/or analysis of the effect of the actual fault on the measurements selected. For example, the data shown in Figures 10.4 and 10.5 describe the performance of a turbine which has sustained damage. The physical damage is to the geometry of the blades; however, there is no practical means of making such a measurement in flight. It must be inferred from performance measurements that can be obtained during normal operations. Such a state of affairs suggests that a thermodynamic model of the engine in which the damage mode has been included would prove a very useful tool in the assessment of the accuracy; a selected set of performance measurements could represent the actual fault.

Regardless of the engineering methodology used to assess accuracy, the measurement suite must be able to indicate the fault and make a determination of its severity. If this cannot be obtained, the particular measurement suite must be abandoned in favor of another.

Also shown in Figure 10.14 is the requirement to test for ambiguity, which is frequently more difficult to ensure than accuracy. This is especially so for performance measurements or for those involving complex signal analysis. In either case, the question is the same. Are there other modes of failure that could excite the same set of measurements in exactly the same way? It is emphasized that the problem usually has many dimensions and the suite of measurements is always as small as is practical. Where possible, investigations must therefore address the issue of ambiguity. Where ambiguity exists, it can frequently be dealt with by the inclusion of one or more additional measurements. This is particularly true of performance measurements where both the direction of change (plus/minus) relative to a baseline together with the relative magnitude of the change can isolate a specific problem.

At this point, it is perhaps helpful to discuss the nature of the fault library and its use. Notionally, a fault library is a matrix of data where the rows are measurements and the columns are faults, as indicated in Figure 10.15.

In this figure a particular fault is characterized by a number of measurement parameters in which a direction of change away from a baseline is indicated. For example, a parameter may be an exhaust gas temperature measurement corrected by inlet temperature and compared to a baseline of this parameter plotted as a function of corrected rotor speed. Such a baseline is shown in Figure 10.16.

As shown in the figure, a measurement must first be corrected to local inlet temperature and then compared to the baseline at the corrected rotor speed. The deviation obtained forms an element in the fault matrix which, in the example shown, indicates a positive deviation. Should this parameter be represented in the fault library matrix as P_{m1}, it

Measured parameter	Fault				
	F_1	F_2	F_3	F_4	F_5
P_{m1}	↑	↓	↓	—	↓
P_{m2}	↓	↓	↑	↑	—
P_{m3}	—	↑	↓	↑	
P_{m4}	↑	↑	↑		
P_{m5}	↓	↓	—		

Figure 10.15   The fault library matrix.

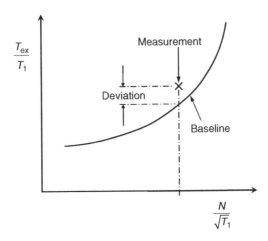

Figure 10.16   Typical baseline.

would indicate a positive change and (in the example indicated in Figure 10.15) it would be associated with the fault F_1.

As can be deduced from this discussion, the purpose of the fault matrix as part of the fault library is to resolve possible ambiguity and to allow the fault to be isolated to a single cause. Other elements of a working fault library will allow limits to be imposed on the magnitude of the deviations before warnings and alarms are annunciated. In a fully developed condition-based maintenance system, ground personnel would be informed of the condition and the necessary actions organized to minimize losses of availability of the machine.

Returning now to Figure 10.14, the remaining task in the qualification of a fault indicator is the issue of precision. This is a practical application of statistics aimed at automating the processes involved in implementation. It applies to all measurements that have been included in any fault indicator. The essential requirement of this test is that the measurement be statistically well behaved across not only the measurements taken from a single machine but also across a fleet of the same machines. Figure 10.17 is a histogram of a particular fault indicator for an entire fleet of engines taken over about 2500 flights. In this particular case, the parameter is the familiar w_F/P_C upon which the fuel limit schedule is built.

As can be observed from the figure, the statistical character of the data is quite well behaved. A single clear maximum exists and the bulk of the measurements fall within reasonable bounds. Such a state of affairs allows the establishment of a minimum and maximum limit which can be applied across the entire fleet. Recalling that low values of this parameter would suggest sluggish dynamic response and therefore long acceleration times, the validity of the measurement is readily checked at the lower edges of the plot. However, high values of w_F/P_C suggest the likelihood of in-flight stalls. A maximum allowable value of w_F/P_C can therefore be established and used to trigger maintenance actions.

Between the minimum and the maximum values of this parameter, the measurements can be trended as a function of time. Furthermore, this parameter proves to be a useful indicator of the state of the engine and its control at the time of installation.

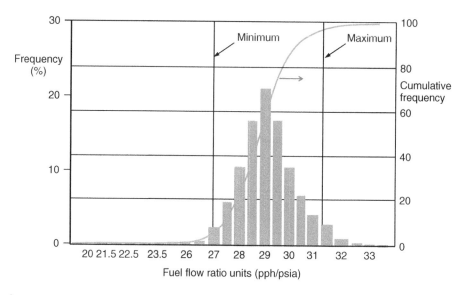

Figure 10.17   Fault indicator qualification.

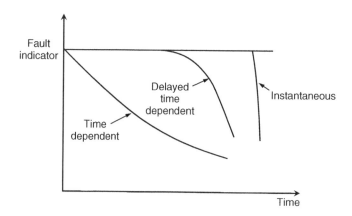

Figure 10.18   Fault indication versus time.

10.3.3   The Element of Time in Diagnostics

All machine failures have some relationship to time. The nature of this relationship has a significant impact on the practical problems of data collection and analysis in any PHM program. In general, this relationship is defined by the change in the fault indicator as a function of time. Figure 10.18 is indicative of the situation.

This figure is a highly generalized depiction of how a fault indication may change with time; however, it serves to highlight that there are several recognizable categories of failure which may occur in a practical situation. The first of these is a purely time-dependent failure where the indication of change begins the moment the part or element enters service and continues at some undefined rate until a limit is reached, at which point the component must be serviced or replaced. An example of such an indicator might be the pressure drop across an important filter. The principal unknown in this case is the rate of clogging of the filter which is driven almost exclusively by the presence of particulate in the flow.

The delayed time-dependent type of failure is characterized by no change in the value of the indicator until a certain time has passed and/or an event triggers a change. An example of such a failure mode might be an indicator of metallic debris obtained from a bearing which has suddenly experienced a spall. The spall may have been occasioned by some earlier indicated event; however, since the indicator is a debris sensor/counter, the first indication of distress is the passage of a particle of metal.

The final category of failure is shown in Figure 10.18 as instantaneous. In this situation the indicator shows no change until the event occurs, at which time there is an instantaneous change in the fault indicator. An example of such a failure mode might be the abrupt failure of a temperature sensor occasioned by mechanical damage or a failed connection.

Figure 10.18 is presented without a scale indication. Considering the many possible failure modes in a gas turbine, it is not possible to indicate a scale; however, some general observations are possible.

PHM is founded on the concept of providing sensors for the express purpose of capturing data that can be interpreted as a fault indicator. The rate at which measurements are accumulated is an important factor in ensuring success in preventing secondary damage and/or loss of use. Clearly, if the rate of change is slow enough, manual inspection of sensors will be adequate. It is also clear that an instantaneous failure would not be detected within the interval of sensor polling typical of modern digital electronics.

For those rates of change which are of the order of hours, a fault indicator can be made productive. For instantaneous failures, a different approach is commonly used. A running or rotating log of critical parameters is kept as a function of time with the oldest data discarded as each new dataset is collected. In the event of a sudden change, the log is saved as a record of events. By arranging the data collection to capture a finite period of time before and after the event, the sequence of changes can be analyzed such that a root cause can usually be determined. This type of function is closely associated with controls since the abrupt change of status often dictates an equally abrupt response from the controls.

Diagnosis of such events is usually treated as a postmortem; however, as more sophisticated flight line diagnosis is developed, little time need be lost to assessment of the causes of such incidents.

10.3.4 Data Management Issues

The success of a PHM program is unsurprisingly dependent on a high-quality data management system that will ensure the timeliness of the information and also the association of specific data to the component in question. Figure 10.19 is an information flow diagram indicating the relationship of various data to each other and the functions of the

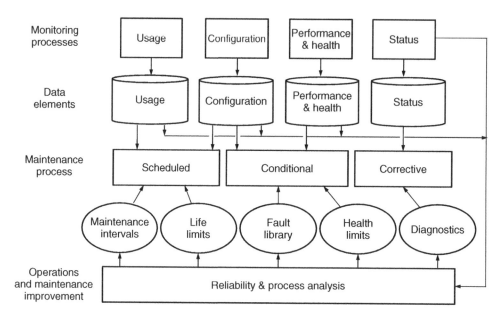

Figure 10.19 Data management of a PHM program.

PHM program. The diagram indicates that PHM is a specific element in an existing over-all fleet management activity where the quality and timeliness of information is of the utmost importance.

The various monitoring processes are indicated across the top of the diagram. Of the four categories indicated, the most basic is the need to maintain strict control of the configuration of the machine. As discussed previously, there are several levels of main-tenance activities present (flight line, operating base, R&O shops), each of which engage in the replacement of parts. Furthermore, there are a number of parts subject to LCF for which the lives of the parts are dictated by law. These parts drive shop visit rates. More importantly, they result in a continuous scrambling of parts on a specific machine as components emerge from an R&O process and become available for re-installation. Such a state of affairs can only be managed with rigorous tracking of parts movements and configuration of complete engines.

Once installed in an aircraft, the rate of usage must be tracked and assigned to the specific parts associated with that configuration. As discussed previously, modern engines relate life consumption to the severity of duty. Life consumption algorithms are therefore complex functions of measured speeds, pressures, and temperatures. Calculation of life remaining and its correlation to aircraft mission is fundamental to the assignment of aircraft and the more general problem of coordination of aircraft availability.

The availability of modern computers allows the development of interactive software which greatly facilitates the handling of data used in the monitoring of jet engines. The heart of this software is a relational database in which the individual components are recognized in terms of their attachment to an engine and/or their location in an R&O pipeline. Figure 10.20 is a screen image of such a configuration manager in a software

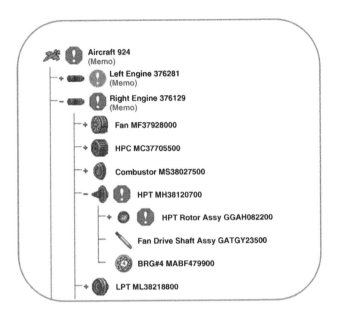

Figure 10.20   Screen image of a parts system software manager (courtesy of GasTOPS Ltd).

example. This illustration provides the reader with a sense of identification and location of a specific part.

As suggested by the image in Figure 10.20, Aircraft 924 is fitted with engine 376281 which, in turn, is comprised of major modules of fan, HP-compressor, combustor, and so on. Further, the fan module can be expanded to the fan rotor assembly, bearing, and fan casing assembly. Not shown, but indicated in the image by a positive sign, the fan rotor can be further expanded to a disc and a set of blades, each of which is identified by a serial number.

The above configuration manager is associated with data files for each of the recognizable components. It is therefore straightforward to associate usage data as per Figure 10.19 with the engine and, through the relational database, link it to individual subassemblies and parts of the system. It is equally straightforward to collect performance and health data and 'attach' these data to the component or part involved. If the component is removed from the engine, the complete data history can move with it. All work done by one or another of the repair pipelines becomes part of the record for that module or part. If it is then re-installed in a different engine, all data remain intact and all data associated with this new assignment is added to the records for the part in question.

Referring again to Figure 10.18 and bearing in mind that a data management system is employed, it becomes obvious that a complete operations and maintenance plan will have defined all of the inputs for each element of the overall O&M activity. Considering the scheduled maintenance component of Figure 10.19, the usage data assigned to specific components can be automatically compared to established life limits; maintenance established by time and/or usage intervals can be anticipated and triggered as required. Similarly, performance and health data obtained from measurements can be trended as a function of time and, as the data approach assigned limits, they are compared to a fault

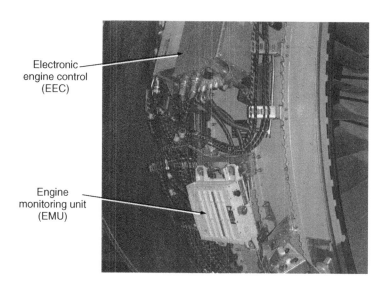

Electronic
engine control
(EEC)

Engine
monitoring unit
(EMU)

Figure 10.21   Fan-case mounted engine monitoring unit (EMU) (courtesy of Meggitt plc).

library and the fault identified. The results of these analyses and comparisons can then trigger specific maintenance actions which are made ready for execution at some time convenient to operations. Finally, where corrective action is required at the flight line, such a system can greatly assist in the diagnostic process.

The final element identified in Figure 10.20 is the 'continuous improvement' component associated with reliability and process analysis. This is an off-line activity which uses the data collected from the on-board monitoring, as well as inspection and repair records, to extend parts lives, improve and extend the fault library, and extend or improve health limits.

A very current example of an engine-mounted engine monitoring unit (EMU) by the Meggitt Plc is shown in the photograph of Figure 10.21. This is typical of the type of modern EMU hardware that supports such engines as the Rolls-Royce Trent, GEnX, and GP7200 series of engines available on the Airbus A380 as well as the new Boeing 787 and 747-8 aircraft.

While much of the functionality is considered proprietary to the engine manufacture's in question, this unit typically supports the following engine condition monitoring (ECM) functions:

1. recording and transmitting accumulated parts life consumption on all life-limited parts of the engine;
2. the monitoring and recording of all designated engine performance variables for analysis by the engine manufacturer (together with the electronic engine control or EEC);
3. processing sensor data and other information available from the EEC and executing a variety of engine diagnostic algorithms for rapid transmission via aircraft communication and reporting system (ACARS) to operator maintenance personnel and subsequently to the specialists within the engine manufacturer organization; and

4. recording and processing of flight-critical engine parameters specified by the regulatory authorities and transmitting them for display in the cockpit;

5. functional integration with the EEC via dedicated data bus.

The unit communicates with the aircraft via ARINC 664 and provides processed and/or raw data to a variety of communication systems which convey the information to ground stations. ARINC 664 is an avionics full-duplex switched ethernet (AFDX) data network for safety-critical high-bandwidth applications.

An avionics bay-mounted unit having similar functions is available for the Boeing 777 aircraft.

References

1. NTSB (1996) Uncontained Engine Failure, Delta Airlines Flight 1288, McDonnell Douglas MD-88, N927DA, Pensacola, Florida, July 1996, NTSB/AAR-98/01, PB98-91041, Adopted January 13th, 1998.
2. Lipson, C. and Sheth, N.J. (1973) *Statistical Design and Analysis of Engineering Experiments*, McGraw-Hill Publishing Company.
3. Tobias, P.A. and Trindale, D.C. (1995) *Applied Reliability*, 2nd edn, Van Nostrand Reinhold.
4. Reliability Information Center (2010) Reliability Modeling – The RIAC Guide to Reliability Prediction, Assessment and Estimation, Reliability Information Analysis Center, Utica.

11

New and Future Gas Turbine Propulsion System Technologies

Over the past 50 years the jet-powered airplane has established itself as the dominant mode of travel for mankind. Its speed and efficiency in moving goods and people is such that it must now be regarded as basic infrastructure similar to highways and electrical power lines. As such, it is a fairly safe assumption that the airplane and hence the jet engine is here to stay and will continue to be the subject of developments aimed at improving both performance and costs.

The origin of jet propulsion as a wartime initiative is well known and the turbofan, turboprop, and turboshaft engines were predictable developments.

By the late 1950s the civil aircraft industry had advanced to the stage where the turbofan engine had made its debut. The Rolls-Royce Conway was the first such engine, followed closely by the PW JT-3D. Both engines had only modest bypass ratios of the order 0.3–0.6. The turbofan engine necessitates the two-spool arrangement and this step alone presented massive development efforts. Again these engines offered higher power to weight ratios and improved efficiencies. The low-bypass ratio ensured that the engine would be suitable for military applications and much of the basic engineering development was supported by government contracts. Much of this synergism was lost in later years as the civil aircraft industry went to wide-bodied airplanes. The so-called high-bypass-ratio turbofan engine was introduced with bypass ratios greater than 5:1.

This chapter addresses new and future technologies that are either in service as innovative solutions to more traditional gas turbine propulsion system functions, or can be expected to play a significant role in the evolution of gas turbine propulsion systems over the next two decades.

11.1 Thermal Efficiency

The industry continues its relentless pursuit of improved fuel economy and reduced emissions. The latter is largely a matter of combustion system design, including fuel nozzles capable of controlling the primary combustion zone in a manner that ensures complete

Gas Turbine Propulsion Systems, First Edition. Bernie MacIsaac and Roy Langton.
© 2011 John Wiley & Sons, Ltd. Published 2011 by John Wiley & Sons, Ltd.

combustion at the lowest possible primary zone temperature. While such designs can eventually influence the basic engine control, the basic development will be driven by combustion specialists which puts this technology outside the scope of this book.

Improvements in component efficiencies will undoubtedly result in some thermal efficiency improvements; however, this is a fairly well-developed field for which computational fluid dynamics has made major contributions over the past several decades. Further improvements in component efficiencies can be expected to be slower in the coming decades.

The twin parameters of overall pressure ratio and turbine inlet temperature are the principle cycle design parameters for any gas turbine. The addition of compressor stages adds weight and complexity; however, the improvement in overall thermal efficiency is undeniable and the past two decades have resulted in increases in pressure ratio with fewer compressor stages than were previously possible. A historical plot of growth in overall pressure ratio is shown in Figure 11.1.

It is interesting to note that, at a pressure ratio of 50:1, the temperature at the entrance to the combustor is of the order 1275 °F, which approaches the combustor exit temperature achieved by Whittle in 1941.

By far the greatest possibility for improvements in thermal efficiency is the development of new materials which will allow the top turbine temperatures to be raised. A historical plot of turbine entry temperatures is shown in Figure 11.2.

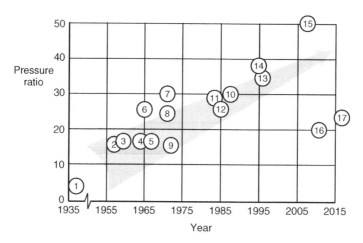

Legend		
1 Whittle	7 GE CF6	13 PW 4084
2 RR Conway	8 PW JT9D	14 GE90
3 PW JT3D	9 TFE731	15 RR Trent1000
4 RR Spey	10 IAE V2500	16 PW 1000G
5 PW JT8D	11 RB211-524	17 CFM LEAP
6 GE TF39	12 PW 2037	

Figure 11.1   Historical growth in overall pressure ratio.

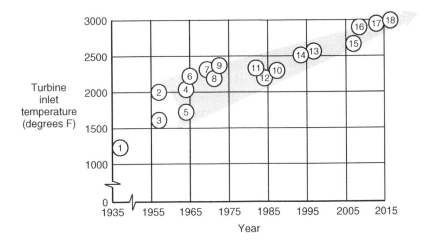

LEGEND		
1 Whittle	7 GE CF6	13 PW 4084
2 RR Conway	8 PW JT9D	14 GE90
3 PW JT3D	9 TFE731	15 RR Trent900
4 RRSpey	10 IAE V2500	16 RR Trent1000
5 PW JT8D	11 RB211-524	17 PW 1000G
6 GE TF39	12 PW 2037	18 CFM LEAP

Figure 11.2   Historical trends in turbine inlet temperatures.

Over the past couple of decades, the US military have funded work under the integrated high-performance turbine engine technology (IHPTET) program. Some of this effort was devoted to the application of ceramic materials to high-temperature applications. The difficulty with ceramics, of course, is their relatively low ductility and therefore their propensity to fail suddenly and catastrophically. Nevertheless, this technology has advanced to the point where it is worth serious consideration for non-rotating components such as nozzle vanes and liners. It has recently been demonstrated on rotating turbine blades for a military application [1]; however, it is thought to be several years away from consideration for a civil engine application. In all cases, the materials being developed offer reductions in weight. For example, the adoption of ceramic matrix composites suggests a reduction of 30% in the entire turbine system if the material proves capable of survival in the duty typical of gas turbine rotating components.

How would such developments effect engines from a systems perspective? Success is somewhat dependent on the ability to control the uniformity of temperature distribution at the entrance to the turbine. This could force the use of more accurate and sophisticated temperature measurement systems which, in turn, will demand a more distributed control of the combustion process. This suggests that much work has yet to be done before these new materials technologies are brought into a real-life civil/commercial service application.

11.2   Improvements in Propulsive Efficiency

Within the next decade, the commercial aircraft industry is expected to replace the highly successful B737 and A320 series of single-aisle aircraft with next-generation more efficient products. In a similar time frame, replacements for the B777 and A330 twin-aisle aircraft are also being considered. For this to make business sense, the fuel efficiency, acquisition, and life cycle costs must be demonstrably much improved over the present generation of aircraft. In support of this objective, the major engine manufacturers are expected to be able to offer new propulsion system solutions with at least 15% improvement in specific fuel consumption over current in-service designs. This may be obtained by straightforward improvements in thermal efficiencies through higher temperature and pressure engines, improvements in propulsive efficiency, or some combination of the two.

For high-bypass engines we can make the reasonable assumption that the nozzles are unchoked so the thrust is defined as

$$F = w(V_j - V_{a/c})$$ (11.1)

where w is engine mass flow, V_j is velocity of the jet, and $V_{a/c}$ is forward velocity of the aircraft. For a given aircraft speed, we can therefore get the same thrust from an engine of high mass flow and low jet velocity as we can from an engine with small mass flow and high jet velocity. Since the kinetic energy left in the jet after it leaves the engine is essentially lost, it is obvious that a lower jet velocity offers a higher efficiency. Indeed, propulsive efficiency η_p is defined by Saravanamuttoo *et al.* [2] as:

$$\eta_p = \frac{2}{1 + V_j/V_{a/c}}.$$ (11.2)

Taken to its limit, when $V_j = V_{a/c}$ the propulsive efficiency $\eta_p = 100\%$; however, the thrust at this point is zero. This observation argues, therefore, for V_j to move towards $V_{a/c}$ as much as is practical. In design terms, this translates to higher mass flow and lower jet velocity which, in turn, means a larger diameter fan and higher bypass ratio.

The practical reality of high bypass ratio is that it demands a lower fan rotational speed partly to keep blade stresses under control and partly to keep blade tip speeds at or near sonic conditions. Since the fan is driven by the LP turbine, this unit is forced to get bigger both in diameter and in the number of stages required to achieve the lower speeds with reasonable efficiency. It is therefore no surprise that the bypass ratio has been set to about 5:1 for most turbofan engines since the 1970s. A plot of the growth in bypass ratio is shown in Figure 11.3.

The quest for improved propulsive efficiency has taken several paths over the past three decades. The main initiatives have been:

1. a gearbox between the LP turbine and the fan to provide the needed speed match between turbine and fan; and
2. a redesigned turbine without stators; each set of blades comprises a stage rotating in the opposite direction to its adjacent one.

Option (2) above effectively halves the rotational speed of the turbine compared to a conventional design. Such a design forces the use of a two-stage counter-rotating fan.

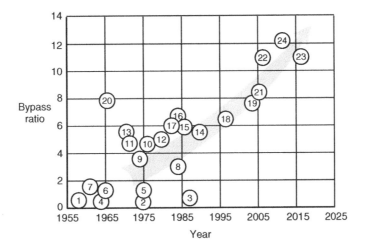

LEGEND			
1 RR Conway	7 PW JT3D	13 PW JT9D	19 RR Trent900
2 GE F404	8 RR Tay	14 IAE V2500	20 GE TF39
3 EJ200	9 TFE331	15 CFM56	21 GE90
4 RR Spey	10 RR RB211-584	16 PW2037	22 RR Trent1000
5 PW F100	11 RR RB211	17 GE CF6	23 CFM LEAP-X
6 PW JT8D	12 GE CF34	18 PW4077	24 PW1000G

Figure 11.3   Historical growth in engine bypass ratio.

Both concepts have also been explored during the 'open rotor' or 'unducted fan' concept demonstration program. These developments were initiated during the 1970s in response to the OPEC oil embargo. However, the price of oil had declined sharply by 1987 and the original efforts were abandoned. Pratt & Whitney however continued development of high-power light-weight gearboxes and has recently introduced a new engine which utilizes a geared fan along with a number of other improvements. In response to this potential competition, General Electric has offered a new higher bypass ratio engine with a combination of improvements in both thermal and propulsive efficiency. These two new engine developments are addressed in more detail in the following sections.

11.2.1   *The Pratt & Whitney PW1000G Geared Turbofan Engine*

As already discussed, previous generation turbofans have been operating with bypass ratios in the vicinity of 5:1. This limitation is largely due to the speed mismatch between the LP turbine and the fan which it drives. A higher bypass ratio favors a slower rotational speed for the fan. At current design values of 5:1, the blading is supersonic in the tip section; increasing the bypass ratio without a decrease in rotational speed means higher stresses and, very likely, a loss of efficiency. At the same time, lower speeds for the LP turbine means a loss of efficiency or an unacceptably large number of stages.

The Pratt & Whitney Geared Turbofan has its origins in the 1980s when NASA funded the development of a 'propfan' demonstrator. This engine, designated the 578DX, was undertaken as a joint effort between Pratt & Whitney and Detroit Diesel Allison. Since Allison had many years experience with airborne gearboxes, the design utilized a gearbox between the engine and a sophisticated counter-rotating variable pitch propfan. The propfan had two sets of six-bladed propellers with substantial sweep back and curvature. It was said to have the speed and thrust of a turbofan and the fuel economy of an advanced turboprop.

The engine made its first flight fitted to a modified McDonnell-Douglas MD-80 in 1989. However, the price of oil dropped at that time and the industry lost interest. Subsequent to the 578DX program, Pratt & Whitney continued its interest in the geared turbofan. However, the problems exposed by the 578DX program (most notably the difficulties of integration with the aircraft and other problems of noise, weight, and reliability) suggested that a high-bypass-ratio conventional fan was a more practical choice. Over the past 20 years Pratt & Whitney has continued with research into reliable, lightweight gearboxes. This work has culminated in a gearbox for the PW1000G as shown in Figure 11.4.

Starting in 1998, Pratt & Whitney built a number of geared turbofan engines each introducing new technologies and improving on previous designs. An early version of this design is shown in Figure 11.5. At that point, the HP compressor comprised five stages; this has since been increased to eight stages which provides considerable improvement in thermal efficiency.

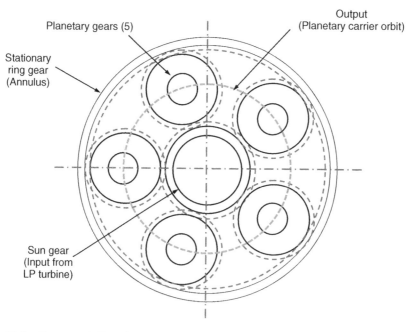

Figure 11.4 Conceptual Drawing of the PW1000G Epicyclic Gearbox (courtesy of Pratt & Whitney).

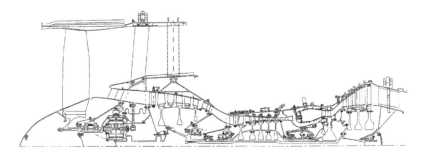

Figure 11.5   Cross-section of the PW1000G engine (courtesy of Pratt & Whitney).

In July 2008, this program was renamed the PW1000G 'Pure power' engine and offered to the marketplace. So far, this engine has found application on several major programs:

1. Bombardier C-Series;
2. Mitsubishi Regional Jet;
3. Irkut MS-21; and
4. Airbus 320 New Engine Option (NEO).

This engine has a bypass ratio of 12:1 and develops thrust in the range of 15 000–30 000 lb with fan diameters ranging from 56 to 81 inches, depending on the thrust level.

In terms of technological improvements, the engine needs only three stages on the LP turbine compared to between five and seven stages for conventional designs. This translates to 1500 fewer parts and, as a result, is expected to yield significant weight savings and an attendant decrease in maintenance costs.

The HP compressor is fitted with single-piece blade/disk (blisks) for each rotating stage. The combustor liners are so-called floating walls which are free to expand and contract independently; the combustion process exploited is a so-called rich/quench/lean cycle under precise control. This technology is the result of a long development effort whereby the flame is controlled to as near the stoichiometric limit as possible in the rich part of the cycle, and is then quenched and run lean to reduce the amount of nitrous oxides (NO_x) produced. Production of NO_x is strongly dependent on flame temperature; such a design reduces the average temperature in the primary zone thus reducing NO_x emissions. Overall, Pratt & Whitney claim to have reduced emissions by 70% which is in line with the recommendations of the International Civil Aviation Organization (ICAO) Committee on Aviation Environmental Protection [3].

Because the fan rotates at one-third the speed of the LP turbine, noise is reduced by 15 dB and the fuel savings have been demonstrated at 12% below conventional existing engines. A new and interesting feature of this design is the use of a variable area fan nozzle. The nozzle is fully open at takeoff conditions to obtain the maximum thrust possible and, as the engine transitions to cruise, the nozzle is closed by as much as 15% to increase fan pressure ratio and reduce fuel consumption. The gearbox is rated in the range of 30 000–35 000 SHP and claims an efficiency of 99%. By gear design standards this is very good indeed; however, it still represents several hundred horsepower in the form of heat which has to be removed. Furthermore, to achieve high reliability, this heat must be removed from the matching gear surfaces which, in turn demands a very advanced lubrication delivery and cooling system. Little has been published on this aspect of the PW100G design; however, there is no doubt that heat removal will be the key to reliability in service.

Pratt & Whitney are silent on their utilization of advanced materials for the engine hot section; however, they have disclosed that the engine fan blades utilize a so-called hybrid metal structure which is considered a major trade secret by the company. Similarly they claim that, through continued development, they can improve the fuel economy to a 22% improvement over existing engines within 10 years. Again they are silent on how this is to be achieved; however, we can readily speculate that thermal efficiency will be the target of this effort and advanced materials will play a key role.

Finally, Pratt & Whitney are predicting a 20% reduction in maintenance costs. The smaller parts count is central to this claim, although the industry considers the introduction of a 30 000 HP gearbox to be a major gamble on reliability.

11.2.2   The CFM International Leap Engine

The so-called Leap engine from CFM International, a very successful joint venture between General Electric and Société Nationale d'Étude et de Construction de Moteurs d'Aviation (SNECMA), has yet to run in a target specific configuration. However, this company dominates the single-aisle aircraft market with its CFM56 product. It is known to be working on a number of technologies that could provide the market with a new offering that would be attractive enough to hold their dominant position as an engine supplier.

Since the Leap engine is not due to run until 2014, hard and reliable information on its design and performance is difficult to obtain. However, it is apparent that CFM International intends to build a new engine with substantial improvements in thermal efficiency and some improvements in propulsive efficiency due to the implementation of a higher bypass ratio. The engine has thus far been selected for application on the new Chinese COMAC 919 which is a single aisle aircraft that could compete in the B737/A320 market. The Leap engine is also being offered by Airbus on its NEO (new engine option) version of the A320 series of aircraft.

For a given turbine inlet temperature, specific fuel consumption can be improved with higher core pressure ratios. The Leap engine is therefore intended to operate at a pressure ratio of 22 compared to 12 for the existing CFM56. Such a design requires a 10-stage HP compressor and a 2-stage HP turbine. The CFM56 operated with a single-stage HP turbine which was an advantage in terms of reliability and maintenance costs.

So far, CFM International are silent on the issue of turbine inlet temperature, but General Electric (who supply the core engine) is known to have been working on advanced materials for several decades. Their ceramic matrix composite (CMC) is capable of handling higher temperatures than conventional super alloys; however, it would appear that GE is reluctant to commit to their use in rotating components. Their use for stationary components seems likely, suggesting some rise in turbine inlet temperature.

Also supporting a rise in turbine inlet temperature is the introduction of new developments in blade and vane cooling. This is also a technology that GE continues to pursue and will likely appear in this engine.

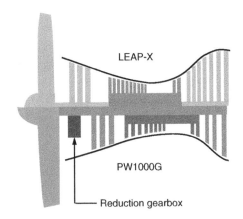

Figure 11.6   Comparative layout of the Leap-X and the PW 1000G engines.

The other major area of improvement in fuel consumption is the proposed use of a fiber composite fan with a bypass ratio of 10:1 compared to between 5:1 and 6:1 for the various versions of the CFM56. The fiber composite design uses a manufacturing technique referred to as 3-D woven transfer molding. This apparently provides the strength, stiffness, and durability required for such an application. This development is well advanced and has been demonstrated by SNECMA within the last year.

The bypass ratio of 10:1 suggests that the larger fan will have to run slower to accommodate possible aerodynamic losses. This will affect the design of the LP turbine, making it larger. In addition, it will almost certainly require more stages because of the lower speeds demanded by the fan. What information is available at this time suggests that the LP turbine will have seven stages: however, the final design has not yet been agreed upon.

As with its competitor, the Leap engine will utilize a 'lean burn' combustor technology to reduce emissions and claims a 50% reduction in NO_x. Its claim of 16% reduction in CO_2 is directly related to its claim of a 16% reduction in fuel consumption. A comparative layout of the Leap-X and the PW1000G, based on the latest publicized information, is shown in Figure 11.6.

This figure shows the tradeoff in relative size and number of stages of the LP turbine for each engine against the gearbox on the PW1000G.

11.2.3 The Propfan Concept

As discussed previously, NASA has done considerable work on advanced propeller technology which produced the Pratt & Whitney/Allison 578DX demonstrator engine. The principle technological advancement was in the higher speed swept, curved propeller blade which substantially increased the disc loading (thrust) at the same time, keeping losses to a low value.

The high swirl of such a propeller suggested the use of counter-rotating propellers. The combination of advancements was therefore named the 'propfan'. It is also referred to in the literature as the 'open rotor concept'. As discussed previously, the 578DX used a gearbox to provide the speed match and the counter rotation needed.

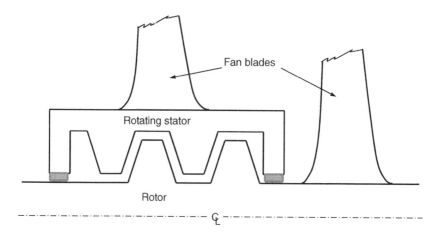

Figure 11.7   Counter-rotating turbine concept used on the GE-36.

In the same time period, General Electric invested in its own propfan demonstrator program, producing the GE-36 which made its first flight on a Boeing 727 test bed in 1986. It is emphasized that this engine was a technology demonstrator and not a commercial engine. As the cost of oil subsided in the late 1980s and the CFM-56 engine became a dominant player in the single-aisle commercial market, General Electric lost interest and the program was terminated.

Perhaps the most interesting feature of the GE-36 was the method used to drive the counter-rotating fans. GE built a counter-rotating LP turbine whereby both stators and blades rotated as sketched in Figure 11.7.

The mechanical design of this arrangement was non-trivial. No gearbox was used so the turbine had to be matched to the propeller insofar as rotational speed was concerned. The pitch control mechanism had to be fitted around the rotating barrel formed by the rotating 'stators'. A photograph of the GE-36 is shown in Figure 11.8.

Both the 578DX and the GE-36 were able to demonstrate substantial fuel savings (of the order 30%). These accomplishments were made without any of the advancements in thermal efficiency that have occurred over the ensuing 25 years. Nevertheless, both engines had problems that would have had to be addressed before they could be considered for commercial applications.

The most obvious problem was noise. The aft propeller chopping through the wake of the forward propeller was a major source of noise and the lack of a shroud exacerbated the problem.

Similarly, integration of the propfan into a civil aircraft was judged to be a potential certification problem. The open rotor posed the ever-present threat of a blade failure. It has been stated that none of the problems identified were or are insurmountable; however, the development costs will no doubt be substantial.

All three major engine manufacturers have indicated that they are considering a revival of the propfan concept, the most consistent of which is Rolls-Royce. Although Rolls-Royce has yet to show its hand in the emerging competition with Pratt & Whitney and CFM International, one of the concepts being studied is the use of a counter-rotating

Figure 11.8 The GE-36 engine (courtesy of Burkhard Domke).

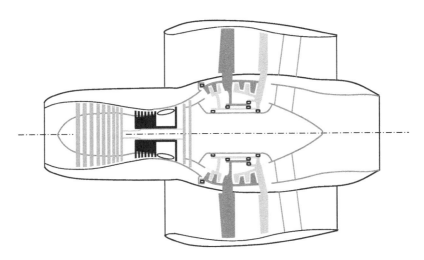

Figure 11.9 The counter-rotating fan concept.

fan/LP turbine as shown in the schematic of Figure 11.9. As shown, the two-spool gas generator powers two counter-rotating shrouded fans, each with a dedicated two-stage free turbine.

The counter-rotation arrangement implies a 2:1 reduction in absolute revolutions per minute for the propulsor rotors. The shaft will be very short and very stiff. There is no suggestion that the fan blades will have variable pitch. This configuration provides greater flexibility to increase bypass ratio and thus improve propulsive efficiency, while at the same time applying the advances in traditional core technologies to increase thermal efficiency. Also, the fan shroud will greatly reduce noise relative to the previous open rotor concepts.

It is worth noting in closing that, in 2008, General Electric signed a 'spare act' agreement with NASA covering studies of the open rotor concept. Under this agreement, the test rigs used in the original demonstration program will be refurbished and new studies conducted into the potential for a future propfan. The inference from this agreement is that the open rotor concept is not dead.

It is worth noting in passing that the gas turbine is a highly developed machine and that each successive development seems to be an increasingly expensive proposition. For example, the counter-rotating propfan will require the simultaneous development of a highly specialized turbine and an equally highly specialized fan. The normal development initiative anticipates a sufficient volume of sales to recover the development costs or the ability to reuse a given technology over a wide range of future engines. While basic technologies such as materials and systems may benefit from military developments, the configurations that apply to military platforms are now substantially different from civil applications. As the civil marketplace continues to further segment into short-haul and long-haul operations and the pressure for lower fuel costs continues to intensify, it is doubtful that any particular segment will be able to support increasingly expensive developments. Such a state of affairs makes some of the more interesting concepts difficult to support, especially where true competition exits.

For more information on the history of the propfan see [4].

11.3 Other Engine Technology Initiatives

So far we have focused on the engine design aspects of technology developments that have occurred and are in process in the pursuit of improved propulsion system performance. The following sections address both the aircraft-level and engine systems-level technologies that have either recently entered into service or are under development. Many of these new technologies that are impacting new propulsion system designs have evolved from the 'all-electric aircraft' initiative launched as far back as the 1970s by the Air-Force Research Laboratory (AFRL) at the Wright-Patterson Air-Force Base Laboratory in Dayton, Ohio. This initiative, driven primarily by the need to simplify the support logistics for military aircraft in theater, later morphed into the 'more-electric aircraft' program which in turn spawned the 'more-electric engine' (MEE). This initiative, in conjunction with the 'power optimized engine' program, continues today supported by government funding and with all of the major engine companies and their tier-one suppliers on both sides of the Atlantic participating.

11.3.1 The Boeing 787 Bleedless Engine Concept

We begin with the Boeing 787 aircraft which, as this book goes to press, is nearing completion of an extensive certification program and is expected to enter service this year (2011).

The power systems architecture on the Boeing 787 is substantially different from any previous commercial aircraft. As mentioned briefly in Chapter 8, the engines which power this aircraft are essentially 'bleedless', meaning that only a minimum amount of bleed air is used to provide engine/cowl anti-icing, hydraulic reservoir pressurization, and oil

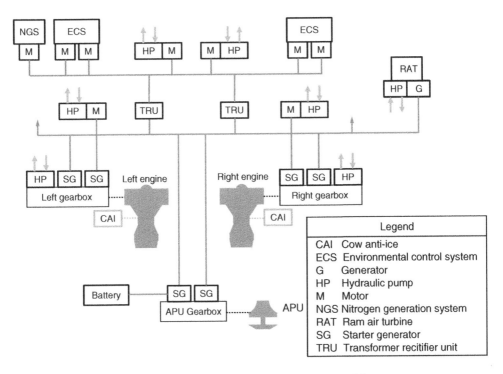

Figure 11.10 The Boeing 787 power systems architecture.

bearing seal pressurization. The anti-icing bleed-air off-take is only required during icing conditions that occur at relatively low altitudes well below the normal cruise regime; the hydraulic reservoir and oil pressurization function do not involve any significant bleed-air flow. Bleed-air usage in this application therefore has a negligible impact on overall engine efficiency. Figure 11.10 is a schematic diagram of the engine power generation system showing the power sources for all of the aircraft control, actuation, and utilities systems.

The benefits of this new architecture include:

- more efficient use of secondary power extraction;
- more efficient engine cycle resulting from the quasi-elimination of engine bleed air as a power source;
- reduced engine build-up (EBU) complexity through the elimination of integrated drive generators (IDGs), pneumatic ducting, pre-coolers, valves, regulators, etc., which have historically been high-maintenance components; and
- elimination of the air compressor and associated ducting from the auxiliary power unit (APU).

The improvement in fuel consumption resulting from the new bleedless architecture is estimated to be about 3%.

As shown in Figure 11.10 there are six starter-generators on the aircraft: two on each main engine and two on the APU. The four engine starter-generators provide 235 VAC (volts alternating current) variable frequency power each with a maximum capacity of 250 kVA. Frequency variation based on the engine speed change from ground idle to maximum HP spool speed is from 360 to 800 Hz. The APU starter generators also provide 235 VAC with a capacity per generator of 225 kVA. APU speed is varied as a function of ambient temperature for optimum APU performance, which results in an APU speed variation (and hence a generated APU power frequency variation) of about 15%.

An additional generator is driven by the ram air turbine (RAT) which can be deployed to provide emergency electrical (and hydraulic) power in the event of loss of both main engines.

The APU can be started by battery power or via power from the ground power receptacle.

When operating in the starting mode, power electric modules (PEMs) facilitate variable speed and variable voltage drive until self-sustaining speed is attained when the starter-generators switch over to the generation mode of operation. Both starter-generators are normally used during engine and APU start; however, a single unit (or ground power) can accomplish a start with an extended start time.

The electrical power system on the 787 aircraft is a hybrid system utilizing a combination of traditional 115 VAC and 28 VDC (volts direct current) together with 235 VAC and ±270 VDC (via the transformer rectifier units (TRUs). While the traditional power buses provide power for the majority of conventional aircraft loads, the high voltage buses support the larger system loads such as air compressors for hydraulic pumps, environmental control system (ECS), and nitrogen generation system (NGS) in order to minimize feeder wire weight.

From the above overview of the 787 power system architecture, it is clear that this aircraft has made big strides toward the definition of the 'more-electric aircraft'. This application, however, is still a long way from the elimination of the engine auxiliary gearbox, which still remains the ultimate goal of the MEE program initiative.

To summarize the 787 application, note the growth in electrical generating capacity of this application with respect to historical norms. Figure 11.11 is a graph comparing the main engine generating capacity for a large section of the commercial transport aircraft relative to the new 787 aircraft. The points in time on the x axis are the year of first flight.

As indicated, the 787 capacity (which does not include the APU generators) is more than twice the capacity of all of the other aircraft with the exception of the A380 super jumbo, which has a generator capacity of 60% of the 787.

The more-electric approach to engine power extraction is a radical change from conventional propulsion systems. This has proved to be challenging in view of the problems encountered with the functional maturity aspects of the power management and distribution system. It remains to be seen if the touted improvements in operating costs can be realized in service.

It is interesting to note that the Airbus A350, which is the competitor to the Boeing 787, has adopted a more traditional approach to engine power extraction with bleed air used as the power source for wing anti-icing, environmental control, cabin pressurization, and fuel tank inerting. The APU also provides compressed air for engine starting.

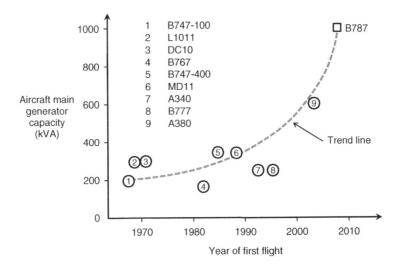

Figure 11.11 Commercial aircraft generator capacity trends.

11.3.2 New Engine Systems Technologies

There are two major areas of technology that continue to dominate the new development and technology demonstration activities associated with engine systems:

- micro-electronics and software (full-authority digital electronic control or FADEC technology); and
- high-power switching technology.

FADEC technology is a mature technology with essentially all new engines using the FADEC for fuel control and thrust/power management. The two-channel dual-dual architecture has become established as the optimum solution, essentially providing 100% fault coverage capability. The in-service engine shutdown rates attributable to the fuel control system have exceeded expectations since the inception of the FADEC in the 1980s. Furthermore, the available memory capacity and throughput provide an almost unlimited functional and monitoring capability. To illustrate this point, more than 20 years ago a FADEC-controlled turbofan technology development engine successfully demonstrated excellent response and handling of a sophisticated control system comprising:

- gas generator fuel control;
- five zones of afterburner control including both core and bypass duct burning; and
- five variable geometry control loops including compressor and turbine geometry and both duct and core exit nozzle area.

Since then the available memory and throughput of the typical FADEC has increased by more than an order of magnitude.

Solid-state power switching technology is the second major enabler of new more-electric initiatives with their ability to switch tens, if not hundreds, of amps at high speed with very low power loss (and hence parasitic heat generation). One important example of this technology is the introduction of new 'power-by-wire' solutions for the primary flight control actuation systems of the F-35 Joint Strike Fighter and the new A380 superjumbo commercial airliner. The F-35 uses electro-hydrostatic actuators (EHAs) on all of its primary flight control surfaces while the A380 employs this same technology (termed electric back-up hydraulic actuators or EBHAs) for control surface actuation redundancy as an alternative to the provision of a third distributed hydraulic system.

From an engine systems perspective, these enabling technologies have made a significant impact on both in-service and technology demonstrator solutions as summarized below.

11.3.2.1   The A380 Electrical Thrust Reverser Actuation System (ETRAS®)

This system, supplied by Honeywell and Hispano-Suiza and available on both the Rolls-Royce Trent 902 engine and the GE/Pratt & Whitney Alliance GP7200 engine, is the first of its kind to enter service on a commercial airliner.

The solution uses electric motor-driven screw-jack actuators in place of the traditional hydraulic or pneumatic actuators to drive the blocker doors. Control electronics and power-switching electronics are installed in separate units within the engine nacelle. Actuator synchronization employs the traditional flexible shaft drive method.

In a related development program, Goodrich has demonstrated a similar thrust reverser approach where the actuators are synchronized electronically (Figure 11.12).

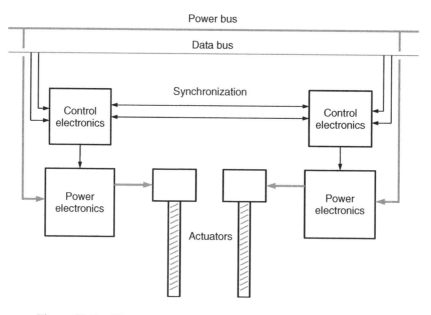

Figure 11.12   Thrust reverser concept with electronic synchronization.

11.3.2.2 Motor-driven Fuel, Hydraulic, and Lubrication Pumps

The ultimate goal of the MEE initiative is the elimination of the accessory gearbox together with the introduction of an electric starter/generator embedded within the engine, driven directly by the engine's HP spool. All of the power extracted is therefore in the form of electrical power with the reverse process used for engine starting.

While it is already common practice to generate aircraft hydraulic power from AC and DC electric motors, the concept of providing high pressure fuel and lubrication oil from an electric power source is a radical change from the current direct mechanical drive methods.

Clearly some level of redundancy is essential in order to ensure comparable levels of reliability, availability, and integrity of these critical functions. Demonstrator programs have utilized dual wound brushless DC motors for this task. Figure 11.13 is a schematic of an electric fuel pumping and metering system (EFPMS) presented at the Teos Forum in 2006 [5]. In this demonstrator the power converter electronics are fuel cooled via the 'fuel return to tank' (FRTT) valve, with the intent that, in practice, this fuel would be discharged back to the aircraft fuel tank.

The benefit of motor-driven pumps for the fuel control is that the motor speed can be optimized to the requirements of the prevailing operating condition. As mentioned in Chapter 3, the traditional gear pump is sized to meet starting fuel flow requirements at about 15% of maximum HP spool speed. This results in a large amount of excess capacity at the altitude cruise condition with the attendant heat generation and loss of operational efficiency.

A real-world example of this new technology approach to fuel metering control is the fuel metering pump control system currently operational on the Boeing 787 APU (see Figure 11.14). This fuel metering control comprises:

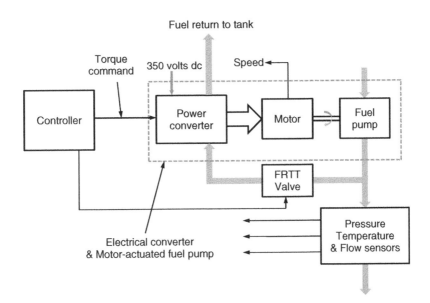

Figure 11.13   Electric fuel pumping and metering demonstrator schematic.

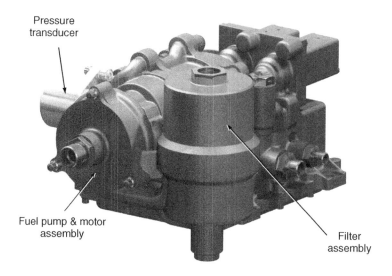

Figure 11.14   The Boeing 787 APU fuel metering pump (courtesy of Parker Aerospace).

- a brushless DC motor-driven fuel pump with both centrifugal and positive displacement gear elements;
- on-board high-power switching electronics for motor commutation, torque, and speed control;
- fuel delivery pressure and fuel filter delta-P sensors; and
- a digital bus interface that provides pump speed commands and transmits feedback of metering unit output parameters and health status to a remote electronic fuel control unit.

The most significant challenge for this concept is to achieve the required in-service reliability, particularly with regard to the electronics and high-power switching devices which reside in a relatively hostile environment.

It is important to note that, in this type of application, the reliability and integrity requirements are substantially less demanding than for a main propulsion engine; this is because the APU is typically only used on the ground and, in many situations, the APU may not be dispatch-critical. Nevertheless, the in-service experience obtained from this application could prove valuable in assessing the technical risk associated with extending this technology to a gas turbine propulsion engine.

The elimination of the accessory gearbox is likely to prove difficult in practice. There is something comforting in realizing that, if the engine is turning, so too is the fuel pump, the lubrication, and scavenge pumps – not to mention the dedicated alternators supplying power to the engine FADEC. With this traditional approach, the functional integrity is clear.

11.3.2.3   The Oil-less Engine

Another initiative that could be a major contributor to the elimination of mechanical power extraction and hence the auxiliary gearbox is the concept of the 'oil-less' engine.

An engine design that is free from the operational problems of managing a lubrication program is very attractive. All gas turbines consume oil at a finite rate; indeed, some would argue that the life of the oil is considerably increased by the refreshment from regular top-ups. As more and more emphasis is placed on unmanned vehicles by military operators, engines can be expected to stay aloft for as much as weeks at a time. This expectation of very long endurance is driving a number of R&D initiatives which are looking into methods of rotor support that do not use conventional rolling element bearings and the lubrication systems that are required for such bearings [6].

An oil-less engine implies that the shafting is levitated by some means which is generated on board, such as high pressure air jets or magnetic fields arranged to maintain the shaft in an orbit central to a journal. Either method would reduce the bearing losses, dramatically suggesting an improvement in efficiency.

The air bearing concept was introduced by Honeywell (formerly Garrett-AiResearch) in the 1960s. Its principal application appears to have been high-speed compressors and APUs, where the problems of managing an oil program were somewhat more severe and the likelihood of foreign object damage was less. Air as a levitating fluid was convenient, since it could be bled from a compressor which was readily available as part of the system.

Currently, there appears to be much more emphasis on the development of magnetic bearings; see [7] for a useful overview of this technology. As with air bearings, the concept is not new. Indeed, the pipeline industry actively pursued this technology throughout the 1980s and 1990s with considerable success [8]. The primary motivation of the pipeline industry was to reduce operating costs both in terms of fuel consumption and maintenance. These pumping units are stationary of course, but distributed over many miles and often in remote locations. In these circumstances, light-duty permanently sealed rolling element bearings were used to deal with relatively infrequent start-ups and shutdowns; however, the magnetic bearings proved to be extremely reliable and reduced bearing losses to nearly zero.

The concept of a magnetic bearing is quite simple. The shaft is surrounded with electromagnetic coils mounted on a stator as shown in Figure 11.15. Sensors determine the radial location of the shaft and the current to the coil of each pole is controlled so that the shaft is maintained in concentric orbit. This sounds simple enough; however, the key is accurate measurement of shaft position and a sophisticated controller capable of computing in real time the necessary current demand of each coil.

The progress made in the last decade in microelectronics makes the concept feasible insofar as functionality goes; however, there are many issues to be resolved before such a system is a working reality. Some of the practical design issues include size, weight, and the method by which the arrangement can be protected in the harsh environment of the hot section of a gas turbine. Indeed, such a design must deal with the losses inherent in generating a magnetic field. Eddy current losses in the stators and electric resistance in the coils result in additional heat in an already hostile environment. Finally, the operational reality of damage to turbomachinery blading due to both foreign objects and thermal and low cycle fatigue (which generally results in substantial shaft imbalances) cannot be ignored. Years of effort have gone into the development of conventional bearings with fluid damped bearing support systems. Some combination of these technologies will likely have to be employed to make magnetic bearings a reality.

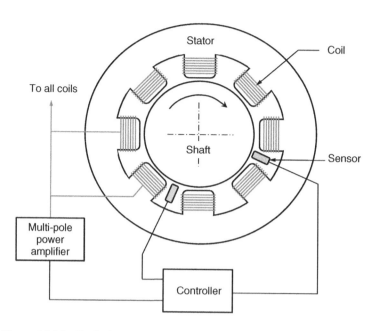

Figure 11.15   Typical arrangements for a radially loaded magnetic bearing.

Perhaps the greatest impetus to further work on magnetic bearings is the idea that they can be adapted to operate as a shaft support *and* as an electric generator. Such an arrangement would certainly support the ultimate MEE goal of elimination of the accessory gearbox, resulting in greatly simplified mechanical arrangements of the engine installation. Such a development is still considered to be a long way from a reality, however.

11.3.2.4   Electro-mechanical Variable Geometry Actuators

As an alternative to the traditional fuel-draulic powered variable geometry actuator, the more-electric approach is to employ a motor-driven electro-mechanical alternative. Demonstration tests so far have used a fully redundant linear actuators concept. It can be argued that the added complexity required to provide a level of reliability and functional integrity comparable with traditional fuel-draulic designs would seem hard to justify in terms of weight and cost; in any more-electric engine solution, high pressure fuel always exists and is therefore readily available to support the actuator function.

11.3.3   Emergency Power Generation

An important issue arising from the more-electric initiatives is worth discussing here as it affects the provision of emergency electrical power in commercial aircraft applications. The traditional solution is the deployable RAT. This device has the serious disadvantage of being inoperative for very long periods of time, to the point where its availability when needed may be questionable. Periodic tests of the deployment and functionality of the RAT are carried out during major maintenance checks but the availability problem remains.

One notable example of what can occur is worth relating. Following a ground check of the RAT, it was found that the relay that switched power from the RAT into the aircraft electrical system had contacts contaminated with a brown sticky substance; as a result, RAT power could not be provided to the aircraft. Further investigation found that the place on the flight deck used by the crew to place their coffee cup was located directly above the RAT relay switching unit, and that coffee was the source of the contamination. This dormant and undetected failure could have resulted in a catastrophic event had the RAT been required in service.

As part of the power optimized aircraft (POA) initiative, successful demonstrations of motor generators suitable for embedding within the engine have occurred. In particular, a switched reluctance generator to be located on the LP spool aft of the LP turbine is considered as a more attractive approach to emergency power generation than the traditional RAT; in this case the generator is in continuous operation and its operational status can be readily monitored, thus circumventing the dormant failure potential of the RAT.

11.3.4   On-board Diagnostics

The role of engine diagnostics is to discover imminent component failures such that safety is not compromised and to support all maintenance-related activities through timely reporting of all in-flight engine degradation. This lofty goal is commonly thwarted by cost considerations. However, even a casual examination of operating costs affirms the large contribution of maintenance to the total. Every failure mode therefore can and should be placed under scrutiny, including the overall concepts of fixed schedule maintenance actions.

The great difficulty with on-board diagnostics is the time and cost required to prove that the diagnostic algorithm is an unambiguous and valid statement of the condition of the engine. Degradation and ultimate machine failure is an unwanted (albeit infrequent) event; a teardown inspection to confirm a diagnosis seldom gets past management scrutiny.

Regardless of the above discussion, the concept of on-board diagnosis is still in its infancy. Much progress has been made over the past two decades, however. The fusion of multiple measurements to ensure that the diagnosis has been qualified is now considered normal, the cost of sensors continues to decline, and operators are becoming accustomed to preparing a ground response to signals forwarded to them from an aircraft. It is expected that this area of technology will continue to grow and be applied more and more widely over the coming years.

References

1. Trimble, S. (2010) Turning up the Heat on CMCs. Flight International, 23–29 November 2010.
2. Saravanamuttoo, H.I.H., Rogers, G.F.C. and Cohen, F. (2001) Propulsive/Overall Efficiency. Prentice Hall.
3. Committee on Aviation Environmental Protection (2007) Environmental Report. ICAO.
4. Sweetman, B. (2005) The Short Happy Story of the Propfan. Air & Space Magazine. September.
5. Teos Forum (2006) Fuel metering demonstrator project by Hispano Suiza. Technology for Energy Optimized Aircraft Equipment and Systems, Paris, France.

6. Sirak, M. (2006) Rolls Royce eyes oil-less engines, other innovative propulsion concepts. Defence Daily, June 8 2006

7. Clark, D.J., Jansen, M.J., and Montague, G.T. (2004) An Overview of Magnetic Bearing Technology for Gas Turbine Engines. NASA/TM- 2004-213177.

8. Alves, P.S. and Alavi, B.M. (1996) Magnetic bearing improvement program at NOVA. ASME International Gas Turbine Conference, UK.

Appendix A

Compressor Stage Performance

This appendix describes the analytical process associated with the stage performance analysis of multistage axial compressors typical of those used in today's modern aircraft gas turbine engines.

In its basic form, the axial compressor is built up by stacking successive stages of rotating blades (rotors) and stationary guide vane assemblies (stators) with the rotor vanes rotating on a common shaft relative to the fixed stator guide vanes.

Since the flow through the compressor is axial, each stage contributes to the compression process which suggests that, if the performance of each stage is known, it should be possible to estimate the overall compressor performance. The resulting analytical process is referred to as 'stage stacking' and has proven to be a useful tool in estimating compressor maps for the purpose of modeling complete engine performance.

The following provides some insight into the origins of compressor stage characteristics, the terminology involved and the generally accepted methods of data presentation. The subsequent appendix (Appendix B) provides the methodology for stacking compressor stage data in order to obtain overall compressor characteristics.

A.1 The Origin of Compressor Stage Characteristics

A compressor stage is a set of rotating blades which impart energy into the flow by changing its angular momentum. The rotating blades are followed by a row of stationary blades (stators) that assist in diffusing the flow and changing its direction such that it is presented appropriately to the next row of rotating blades.

A schematic of a typical compressor stage is shown in Figure A.1. Referring to the above figure, the nomenclature for the associated variables is as follows:

Gas Turbine Propulsion Systems, First Edition. Bernie MacIsaac and Roy Langton.
© 2011 John Wiley & Sons, Ltd. Published 2011 by John Wiley & Sons, Ltd.

v_1	Absolute velocity of the air entering the stage.
v_{r1}	Velocity of the air entering the stage relative to the rotating blade.
U	Tangential blade speed.
α_1	Angle made by the absolute velocity v_1 with respect to the axial reference. Note that this angle is tangential to the trailing edge of the upstream stator. It is established by the orientation of the upstream stator defined by the angle β.
α_2	Angle made by the relative velocity vector with respect to the leading edge of the rotating blade. Notice that v_{r1} will approach the rotating blade at an angle of incidence i with respect to a line tangential to the blade chord.
v_2	Absolute velocity of the air leaving the rotor.
v_{r2}	Velocity of the air leaving the rotor relative to the rotating blade (i.e., tangential to the trailing edge).
α_3	Angle made by the absolute velocity v_2 with respect to the axial reference. Note that this is the angle at which the air enters the downstream rotor.
α_4	Angle made by the relative velocity vector v_{r2} with respect to the trailing edge of the rotating blade.
v_{w1}	Component of the tangential velocity of the incoming air; this is commonly called the 'whirl velocity' at the stage inlet.
v_{w2}	Component of the tangential velocity of the air at the exit of the rotor; this is commonly called the 'whirl velocity' at the stage exit.
v_{a1}	Axial component of the incoming air to the rotor.
v_{a2}	Axial component of the air exiting the rotor.
β	Stator control angle.

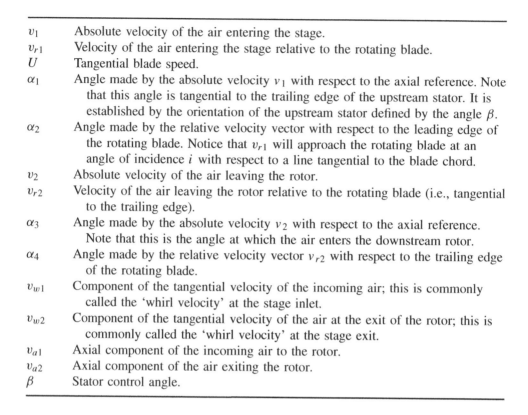

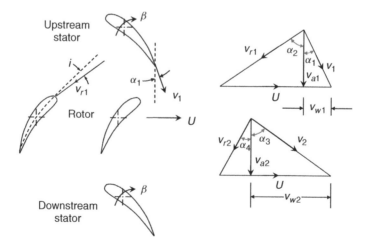

Figure A.1   Axial compressor stage velocities.

It should be noted that the diagram indicates the possibility that the stators can be controlled; in fact, the setting of the stator angle β establishes the value of the air inlet angle α_1, which in turn affects the overall performance of the stage. The effect of β is discussed at the end of the following section.

A.2 Energy Transfer from Rotor to Air

There are a number of valuable references which describe the elementary interactions between a rotor and a fluid [1–3]. The information contained therein is summarized here for the purpose of understanding stage performance.

The torque on the shaft is determined by the change in angular momentum caused by the rotor in accordance with:

$$T = \left[\frac{m_2 r_2 v_{w2}}{t}\right] - \left[\frac{m_1 r_1 v_{w1}}{t}\right] \tag{A.1}$$

where m_1 is mass entering the rotor, m_2 is mass leaving the rotor, r_1 is radius at entry, and r_2 is radius at exit.

Assuming steady-state flow and a constant radius and setting $m/t = w$ (the mass flow rate), the torque becomes:

$$T = wr(v_{w2} - v_{w1}). \tag{A.2}$$

Since power is proportional to torque times angular speed, and observing that the blade speed is determined by $U = \omega r$ where ω is the angular speed, we obtain the expression for power:

$$\text{Power} = wU\,(v_{w2} - v_{w1}). \tag{A.3}$$

From thermodynamic considerations we know that power can also be expressed in terms of the rise in the total enthalpy of the air. We can therefore state that

$$\text{Power} = c_p w(T_2 - T_1) \tag{A.4}$$

where T_1 and T_2 are the absolute temperatures at entry and exit of the rotor, respectively, and c_P is the specific heat of the air.

From Equations A3 and A4, we can relate the temperature rise in the stage to the change in angular momentum as:

$$c_p(T_2 - T_1) = U(v_{w2} - v_{w1}). \tag{A.5}$$

From the velocity diagrams shown in Figure A.1, we observe that:

$$v_{w1} = v_{a1} \tan \alpha_1$$

and

$$v_{w2} = U - v_{a2} \tan \alpha_4.$$

Substituting these expressions for the whirl velocities into Equation A5, we obtain:

$$c_p(T_2 - T_1) = U(U - v_{a2} \tan \alpha_4 - v_{a1} \tan \alpha_1).$$

If we make the assumption that the inlet and exit axial velocity vectors are equal in magnitude, this reduces to:

$$\frac{c_P \Delta T}{U^2} = 1 - \frac{v_a}{U}(\tan \alpha_1 - \tan \alpha_4). \tag{A.6}$$

The term $c_P \Delta T / U^2$ is commonly referred to as the 'temperature rise coefficient' or the 'work coefficient' and is given the symbol ζ.

The term v_a / U is known as the 'flow coefficient' and is given the symbol φ. This is readily deduced from the observation that the flow rate is determined by

$$w = \rho A v_a \tag{A.7}$$

where ρ is the local air density and A is the annulus area of the stage.

From this expression, we can write:

$$\frac{v_a}{U} = \frac{w}{\rho A U}. \tag{A.8}$$

If we now replace ρ with the expression P/RT (basic gas law) we further observe that:

$$\frac{v_a}{U} = \frac{w R T}{A P U}. \tag{A.9}$$

If we now substitute $U = 2\pi N r$ (where N is the rotational speed in revolutions per minute or RPM) we can say:

$$\varphi = \frac{v_a}{U} = \frac{wrt}{2\pi r A P N} = \left[\frac{R}{2\pi r A}\right] \frac{w\sqrt{T}}{P} \frac{\sqrt{T}}{N}. \tag{A.10}$$

For a fixed value of $N/\sqrt{T}$, the flow coefficient is therefore directly related to the non-dimensional flow through the stage.

Returning to Equation A6, we can observe that the expression is linear for given values of α_1 and α_4; a plot of this expression is shown in Figure A.2. The straight-line plot is the ideal or loss-less expression obtained from Equation A6.

The application of frictional and turbulent losses to the flow results in a reduction in the work done coefficient, as represented by the curve. The practical interpretation of this statement is that the turning angles suggested by α_1 and α_4 are not achieved; this is more pronounced at the extremities of the operating range. The pressure across the stage can be deduced from the temperature rise coefficient and the application of an isentropic efficiency. Recognizing that the isentropic efficiency is defined as:

$$\eta_s = \frac{T_2^1 - T_1}{T_2 - T_1} \tag{A.11}$$

where $T_2^1 - T_1$ is the ideal temperature rise.

We can now define:

$$\psi = \frac{c_P(T_2^1 - T_1)}{U^2} = \frac{c_P T_1}{U^2}\left[\left(\frac{P_2}{P_1}\right)^{\frac{\gamma-1}{\gamma}}\right]. \tag{A.12}$$

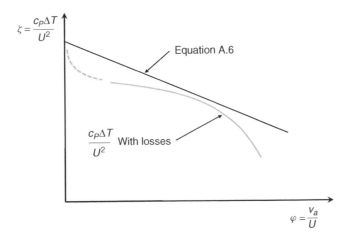

Figure A.2   Idealized stage characteristic.

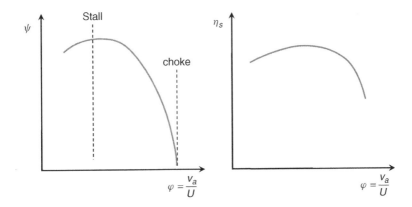

Figure A.3   Compressor stage coefficients.

These parameters together describe the performance of the stage. It is common practice to present stage performance in terms of the pressure coefficient (ideal temperature rise) and efficiency, as shown in Figure A.3.

The figure shows two reference conditions that are worthy of note. Stage stall generally occurs at or near the point of maximum pressure rise. Similarly, the stage is said to be choked as the axial velocity v_a increases to the point where the angle of incidence with respect to the moving blade becomes negative and/or the Mach number approaches the choking condition. In either case, the losses rise to the point where the pressure coefficient falls precipitously.

Returning again to Equation A6 and the diagram of velocity triangles shown in Figure A.1, we can observe that α_4 is essentially controlled by the setting of the rotating blade and α_1 is controlled by the setting of the upstream stator. The provision of angular

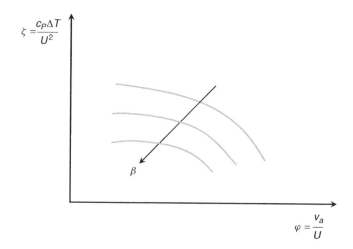

Figure A.4   Variable stator stage performance.

control of the stator (as indicated by the parameter β in Figure A.1) allows external control of α_1 and ultimately the work done on the airflow.

Using β as the control parameter for the stator, it is apparent that increasing β will cause an increase in α_1 which, in turn, causes a corresponding reduction in ζ. This suggests a two-dimensional plot of ζ versus φ for different settings of β, as shown in Figure A.4.

The use of stator control to adjust the work done by a given stage permits a re-matching of flows, temperatures, and pressures such that there is no requirement for interstage bleeds to manage the compressor stall problem during engine acceleration.

References

1. Saravanamuttoo, H.I.H., Rogers, G.F.C., and Cohen, H. (1951–2001) *Gas Turbine Theory*, Pearson Education Ltd.
2. Shepherd, G.D. (1956) *Principles of Turbomachinery*, MacMillan Company, New York.
3. Hesse, W.J. and Munford, N.V.S. Jr. (1964) *Jet Propulsion for Aerospace Applications*, 2nd edn, Pitman Publication Corp., New York.

Appendix B

Estimation of Compressor Maps

For purposes of dynamic analysis of propulsion system performance, a simulation of a jet engine can be a remarkably useful tool. Models which accurately describe the behavior of a gas turbine engine over its full dynamic range require a description of the major components in a form that can be represented on a computer. Such a model is described more fully in Appendix C; however, for purposes of this discussion, it can be stated that the compressor description required is in the form of maps of non-dimensional flow and isentropic efficiency as functions of pressure ratio and non-dimensional speed. A typical compressor map is shown in Figure B.1.

All parameters shown in Figure B.1 are in non-dimensional form, where T_1 and P_1 are the temperature and pressures measured at the compressor inlet and P_2 and T_2 are conditions at exit.

It should come as no surprise that the maps shown in the figure are the result of a long and expensive development effort on the part of the engine supplier. It is also safe to say that the engine performance is, to a great extent, determined by the compressor; in the hands of a knowledgeable engineer, much can be deduced from the compressor maps about the design of the engine. As a result, such maps are highly proprietary and seldom made available to systems analysts. Nevertheless, the power of a good model in performing systems analysis is sufficient motivation to find ways to estimate these maps. This appendix provides a method of representing/estimating the performance of any multistage axial compressor.

The basis of the estimation described here is the establishment of a set of representative stage characteristics and the stacking of the stage data to create an entire compressor map.

The nomenclature used in presenting stage data is defined in Appendix A.

The starting point in compressor map estimation is to gather available data for the engine in question. The following information is typically obtainable with a modest effort of engineering enquiry.

Gas Turbine Propulsion Systems, First Edition. Bernie MacIsaac and Roy Langton.
© 2011 John Wiley & Sons, Ltd. Published 2011 by John Wiley & Sons, Ltd.

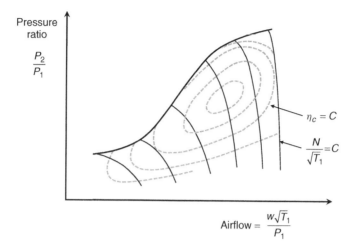

Figure B.1   Typical compressor map.

(a)	Thrust	F
(b)	Specific fuel consumption	SFC
(c)	Turbine inlet temperature	T_3
(d)	Overall compressor pressure ratio	P_2/P_1
(e)	Design point rotor speed	N
(f)	Number of compressor stages	n
(g)	Compressor airflow w or specific thrust	F/w

From the known conditions, we can calculate:

1. fuel flow $w_{\mathrm{Fe}} = F \times \mathrm{SFC}$;
2. temperature rise across the combustor: $T_3 - T_2 = w_{\mathrm{Fe}}\Delta H_f / c_P$ where ΔH_f is the lower calorific value of the fuel and c_P is the specific heat for air;
3. compressor exit temperature: $T_2 = T_3 - (T_3 - T_2)$; and
4. compressor efficiency: $\eta_c = \left[(P_2/P_1)^{\frac{\gamma-1}{\gamma}} - 1 \right] / [(T_2 - T_1)/T_1]$.

From these simple design point calculations, we have a complete definition of the thermodynamic conditions at the compressor design point as follows:

non-dimensional airflow: $\frac{w\sqrt{T_1}}{P_1}$

pressure ratio: $\frac{P_2}{P_1}$

non-dimensional rotor speed: $\frac{N}{\sqrt{T_1}}$

isentropic efficiency: η_c.

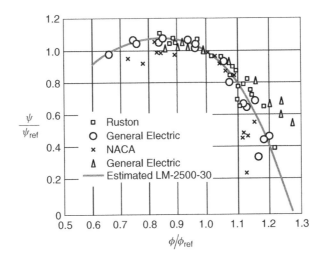

Figure B.2   Generalized stage pressure rise coefficient.

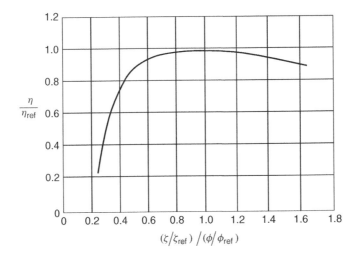

Figure B.3   Generalized stage efficiency.

The fundamental requirement in estimating a compressor characteristic is the availability of stage characteristics. For any specific engine, the stage data will be even more obscure and/or proprietary than the compressor map. However, it is possible to obtain reasonable estimates by scaling non-dimensional data obtained in the open literature. Such data are presented in Figures B.2 and B.3.

Fortunately, a respectable number of researchers have performed experiments on compressor stages and have published their results in the open literature. By collecting and generalizing this data, it can be particularized to a specific compressor design situation.

It has been found that the point of maximum stage efficiency allows the conversion of arbitrary stage performance to be collapsed into a common dataset as shown in Figure B.2.

Considering the pressure rise coefficient for the data represented, it was found that by selecting a reference which represents the point of maximum pressure rise, the data could be collapsed to a single representative curve [1]. As will be shown in Section B.1, the use of these data is built on the assumption that the reference conditions can be interpreted as the design point for each stage in the compressor.

A similar treatment of the stage efficiency is shown in Figure B.3. In this case, the data are collapsed using a somewhat different combination of parameters; however, these data first presented by Howell [2] were found to work quite well for the purposes described here.

Section B.1 particularizes the data of Figures B.2 and B.3 to each of the stages; Section B.2 provides a matching calculation which generates the overall compressor map for all points away from the design.

B.1 Design Point Analysis

The express purpose of the design point analysis is the calculation of the annulus areas of each stage and the computation of the stage performance of each stage at the design point of the compressor.

Obviously, if all aspects of the target compressor are known, there would be no need to make assumptions. This process has evolved because typically less than a complete definition of the machine is available. Thus, several assumptions are necessary as follows.

1. **Blade radius:** it is assumed that a one-dimensional calculation at the mean radius of the blade will suffice. If the value of the mean radii of the blades at each stage is known then, by all means, it should be used. However, in most circumstances it is not known and an assumption of constant radius through the machine will usually produce adequate results. Generally, this radius must be estimated from sketches/diagrams of the front of the engine in question.
2. **Flow area:** the annulus defined by the hub and tip radii of the rotating stage defines the flow area for a given stage. Again, it would be helpful to know the flow area for each stage. In the absence of such data, it will be necessary to estimate the flow area at entrances to the compressor. A sketch of the front of the engine and/or compressor together with a single dimension such as the overall diameter will allow the scaling of the sketch to obtain estimates of hub and tip radii at the first stage. This data is adequate for a first estimate of compressor performance. The mean radius of the blade passage follows from this data.
3. **Constant specific heat:** a basic assumption for such estimates is that the specific heat of air at constant pressure does not change through the machine. This is not strictly true, but the variation in specific heat is small compared to the possible variation in stage efficiency and can therefore be considered a secondary effect.

The above assumptions will allow us to proceed with the design point analysis, beginning with the calculation of the stage flow coefficient and isentropic temperature rise across the whole compressor as follows.

Stage flow coefficient is defined

$$\varphi_d = R \left(\frac{60}{2\pi r A} \right) \left[\left(\frac{w\sqrt{T_1}}{P_1} \right) / \left(\frac{N}{\sqrt{T_1}} \right) \right]$$

(B.1)

where r is the mean blade radius, A is the annulus flow area, and R is the universal gas constant. The value of φ_d can be used to access the generalized stage characteristic of Figure B.2.

Isentropic temperature rise is defined

$$T_2 - T_1 = T_1 \left[\left(\frac{P_2}{P_1} \right)^{\frac{\gamma-1}{\gamma}} \right].$$

(B.2)

This parameter provides an estimate of the loss-free rise in temperature as shown in the temperature–entropy diagram of Figure B.4.

In reality, there are losses which are described by an efficiency of compression defined by the following equation:

$$\eta_c = \frac{T_2' - T_1}{T_2 - T_1}.$$

(B.3)

The actual temperature rise is therefore determined from:

$$T_2 - T_1 = \frac{1}{\eta_c} (T_2' - T_1).$$

(B.4)

Now the compressor designer must determine the distribution of this temperature rise for each stage. For the purposes of estimating the overall compressor map, we must second-guess what he might do.

The simplest assumption is that the design point temperature rise is equally distributed across each stage. This is unlikely to be exactly correct, but will provide a reasonable

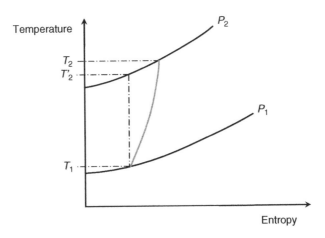

Figure B.4 Temperature–entropy representation of the compression process.

starting point for determining the individual stage characteristics. The individual stage temperature rise is therefore simply:

$$\Delta T_s = \frac{T_2 - T_1}{n} \tag{B.5}$$

where n is the number of stages and the subscript s denotes any stage. From the above assumption, it would be fair to conclude that the isentropic efficiency of each stage is the same as the isentropic efficiency for the entire compressor. Thus, the stage design point parameters are:

$$\eta_{sd} = \eta_c \tag{B.6}$$

and

$$\psi_d = \frac{c_P \Delta T_s'}{U^2} \tag{B.7}$$

where $\Delta T_s' = \eta_{sd} \Delta T_s$ and

$$\zeta_d = \frac{\psi_d}{\eta_{sd}}. \tag{B.8}$$

Applying these design point parameters to Figures B.2 and B.3 we can obtain a set of stage characteristics that will allow stacking of them to estimate the overall compressor map. The only remaining task before the stage stacking process can be undertaken is to calculate the annulus flow area at the entrance to each stage. By fixing the annulus of each stage, we are forcing the stage to operate at the previously determined design point. The calculation is described as follows.

Step 1: Calculate the stage outlet conditions:

$$P_{2s} = P_{1s} \left[1 + \frac{T_{2s}'^1 - T_{1s}}{T_{1s}} \right]^{\frac{\gamma}{\gamma - 1}} \tag{B.9}$$

$$T_{2s} = T_{1s} + \frac{1}{\eta_c}(T_{2s}' - T_{1s}) \tag{B.10}$$

Step 2: Calculate flow parameter at the entrance to the next stage (note that w is a constant):

$$\frac{w\sqrt{T_{2s}}}{P_{2s}} = \frac{w\sqrt{T_{1s}}}{P_{1s}} \left(\frac{P_{1s}}{P_{2s}} \right) \left(\frac{\sqrt{T_{2s}}}{\sqrt{T_{1s}}} \right) \tag{B.11}$$

Step 3: Calculate non-dimensional speed at the entrance to the next stage:

$$\frac{N}{\sqrt{T_{2s}}} = \frac{N}{\sqrt{T_{1s}}} \left(\frac{\sqrt{T_{1s}}}{\sqrt{T_{2s}}} \right) \tag{B.12}$$

Step 4: Calculate annulus area at the entrance to the next stage:

$$A_2 = R \left(\frac{60}{2\pi r \varphi_d} \right) \frac{w\sqrt{T_{2s}}}{P_{2s}} \tag{B.13}$$

Steps 1–4 are repeated for all stages. This process results in a table of annulus areas of the form indicated below:

Stage	Annulus area
1	A_1
2	A_2
3	A_3
n	A_n

B.2 Stage Stacking Analysis

Once a set of stage characteristics have been determined and the annulus area for each stage computed, it is possible to match the stages to each other to obtain an overall compressor operating point. These calculations are usually iterative because they are constrained by the fact that every stage operating point must be located on its own characteristic. The range of a specific stage operation is limited by choking conditions at high values of the flow coefficient $\varphi = v_a/U$ (see Appendix A) and surge or stall at low values.

The stage-stacking calculation is reasonably simple and proceeds as follows.

1. Select a rotational speed N with the intent to compute values of non-dimensional flow, pressure ratio, and overall efficiency that represent valid points on a given 'corrected speed line'. The compressor map shown in Figure B.1 comprises individual curves for constant values of $N/\sqrt{T_1}$. These individual curves or 'speed lines' determine the calculations that will be performed for constant standard inlet conditions of T_1 and P_1.
2. Select an arbitrary value of non-dimensional flow $w\sqrt{T_1}/P_1$. Since many values of non-dimensional flow will result in invalid results, it is helpful to restrain the guess to reasonable values. Here experience and knowledge of the typical range of this parameter at various non-dimensional speeds is very helpful. For example, a typical plot of non-dimensional flow versus non-dimensional speed is shown in Figure B.5 for a typical aircraft engine compressor. This data is representative and is presented as percentages of design point conditions. A number of observations about this plot are relevant to this discussion.
   a. At the upper end of the speed range, the values of maximum and minimum flows converge. Under these conditions, the rear stages become choked and a single value of non-dimensional airflow $w\sqrt{T_1}/P_1$ may be all that is possible.
   b. At the lower speeds, the range of flows is wider indicating flatter speed lines. We also expect that, under low-speed conditions, the front stages are operating close to a stalled condition. This will likely be confirmed by our calculations.

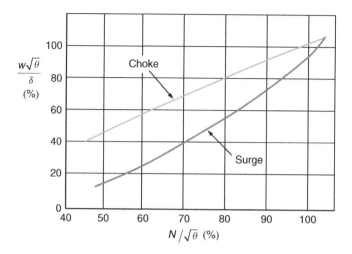

Figure B.5 Typical flow range of a turbojet.

   c. The lowest speeds expressed in non-dimensional terms are of the order 50% of
design maximum under the best of circumstances. The effort to get the compressor
to run at idle conditions of the engine is not insignificant. Thus, a 50% speed line
is representative of the lower limit of most compressor designs.

3. Calculate the value of the flow coefficient for the first stage from:

$$\varphi = \frac{v_a}{U} = R\left(\frac{60}{2\pi r_1 A_1}\right)\frac{w\sqrt{T_1}}{P_1}\left(\frac{N}{\sqrt{T_1}}\right) \tag{B.14}$$

where r_1 is mean blade radius and A_1 is annulus area of the stage.

4. Look up on the stage characteristics the values of ψ and η corresponding to the
calculated value of φ.

5. From these values of stage performance, we can now calculate the rise in temperature
and pressure across the stage as follows:

$$T_2' - T_1 = \frac{\psi_s U^2}{c_P} \tag{B.15}$$

$$T_2 - T_1 = (T_2' - T_1)/\eta_s \tag{B.16}$$

$$\frac{P_2}{P_1} = \left[1 + \frac{(T_2' - T_1)}{T_1}\right]^{\frac{\gamma}{(\gamma-1)}} \tag{B.17}$$

6. The stage outlet conditions determine the conditions at the entrance to the next stage
and are thus determined as follows:

$$T_2 = T_1 + (T_2 - T_1) \tag{B.18}$$

$$P_2 = (P_2/P_1)/P_1 \tag{B.19}$$

It is now possible to calculate the parameters associated with the next stage as:

$$\frac{N}{\sqrt{T_2}} = \frac{N}{\sqrt{T_1}} \left(\frac{\sqrt{T_1}}{\sqrt{T_2}} \right) \tag{B.20}$$

$$\frac{w\sqrt{T_2}}{P_2} = \frac{w\sqrt{T_1}}{P_1} \left(\frac{P_2}{P_1} \right) \left(\frac{\sqrt{T_2}}{\sqrt{T_1}} \right) \tag{B.21}$$

The steps from 2–6 above are repeated for each stage until the operating point of each stage is known and the operating pressure ratio and temperature rise is known for the overall compressor.

By progressively selecting values of $w\sqrt{T_1}/P_1$ which cover the range from surge to choke, the whole speed line can be determined. Also, by increasing the rotational speed and repeating the entire process, each speed line can be established and an estimated performance map for the whole compressor obtained.

The astute reader will observe that some criteria for surge must be applied; this observation is absolutely correct. In general, the criteria of maximum pressure ratio or ψ is used to limit the range of flow that is possible for a given compressor stage. This is especially true for the rear stages. At lower speeds, the front stages can operate in a stalled condition while the overall compressor will still work (albeit at a reduced efficiency). This fact is the origin of such devices as blow-off valves as a means of relieving the front stages. However, at higher speeds it is the rear stages that approach stall first; it is generally conceded that when a rear stage reaches maximum values of ψ, the whole compressor will surge.

A final comment on the calculation process would seem appropriate at this point. As the calculation progresses, it will be found that values of φ at stages downstream of the first stage are likely to be outside the range of the data assumed or estimated for that stage. This simply means that stage is choked and that the value $w\sqrt{T_1}/P_1$ is too high. The calculation must therefore be abandoned in favor of lower values of $w\sqrt{T_1}/P_1$.

References

1. Muir, D.E., Saravanamuttoo, H.I.H., and Marshall, D.J. (1988) Health monitoring of variable geometry gas turbines for the Canadian Navy. *ASME Journal of Engineering for Power*, **111** (2), 244–250.
2. Howell, A.R. and Calvert, W.J. (1978) A new stage stacking technique for axial flow compressor performance prediction. *Transactions of the ASME*, **100**, 698–703.

Appendix C

Thermodynamic Modeling of Gas Turbines

In order to understand and control a complex system such as a gas turbine, it is very helpful to develop a mathematical model of the engine and the systems that support and control it.

There are essentially three modeling methods that are used to support the performance analysis of gas turbine engines.

1. The linear small-perturbation method. Here a linear approximation of the engine functional parameters is developed about a specific operating point. Such an approach to engine modeling is used in the response and stability analyses associated with speed governor performance presented in Chapter 3. This linearized modeling approach is developed fully by Schwarzenbach and Gill [1].
2. A full-range model obtained by extending the linearized approach and modeling the partial derivatives as functions covering the full operating range, to facilitate the study of full-range throttle changes.
3. A component-based approach where models are developed from the aerodynamic, thermodynamic, and mechanical properties of the engine's main modules (compressor, combustor turbine, etc.) and integrated to form a complete model of the engine.

For completeness, all three of the above techniques are described here; however, the main focus is on the third technique listed above which is the most rigorous and complete method of the three.

C.1 Linear Small-perturbation Modeling

The simplest form of linear model is based on a first-order lag where the torque on the engine shaft is cast as a function of speed and fuel flow. From this notion, the instantaneous rate of change of speed can be obtained and the model used to study the engine speed response to small changes in fuel flow.

Gas Turbine Propulsion Systems, First Edition. Bernie MacIsaac and Roy Langton.
© 2011 John Wiley & Sons, Ltd. Published 2011 by John Wiley & Sons, Ltd.

As fuel controls began to evolve, compressor delivery pressure was found to be a useful parameter for the purposes of scheduling fuel. The linear model was therefore extended to describe the dynamic behavior of combustor pressure. Again, a linearized model was developed by recognizing the so-called 'packing lag' associated with changes in airflow during a transient.

C.1.1 Rotor Dynamics

The torque generated on the gas generator rotor can be expressed as

$$Q = f(N, w_{Fe}).$$
(C.1)

Expanding this function using a Taylor Series and neglecting higher order terms yields:

$$\Delta Q = \frac{\partial Q}{\partial N}\Delta N + \frac{\partial Q}{\partial W_{Fe}}\Delta W_{Fe}$$
(C.2)

Considering the acceleration of the rotor shaft, from energy considerations we obtain:

$$J_R \frac{d\Delta N}{dt} = \Delta Q.$$
(C.3)

Combining Equations C2 and C3, we obtain:

$$J_R \frac{d\Delta N}{dt} = \frac{\partial Q}{\partial N}\Delta N + \frac{\partial Q}{\partial W_{Fe}}\Delta W_{Fe}.$$
(C.4)

We can express this in simple block diagram form using the Laplace notation as shown in Figure C.1. The upper diagram is valid for small perturbations about a given operating point; however, if we want to be able to examine the dynamic response at different operating conditions, we must express the variables in non-dimensional terms using δ and $\sqrt{\theta}$ to compensate for variations in air inlet pressure and temperature, respectively. The lower block diagram of the figure expresses Equation C4 in non-dimensional form.

Equation C4 now becomes a simple first-order lag relating fuel flow and speed in the form:

$$\frac{\Delta N}{\Delta w_{Fe}}(s) = \frac{K_e}{(1 + \tau_e s)}.$$
(C.5)

In Equation C5 above, K_e is commonly referred to as the engine gain and τ_e as the engine time constant. These terms are readily calculated as:

$$K_e = \left[\frac{\partial(Q/\delta)}{\partial/(w_{Fe}/\delta\sqrt{\theta})}\right] / \left[\frac{\partial(Q/\delta)}{\partial(N/\sqrt{\theta})}\right]$$
(C.6)

and

$$\tau_e = (J_R\sqrt{\theta}/\delta) / \left[\frac{\partial(Q/\delta)}{\partial(N/\sqrt{\theta})}\right].$$
(C.7)

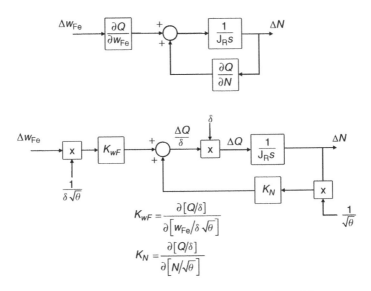

Figure C.1 Linearized block diagram of rotor the dynamics with pressure term.

This formulation provides the response of the rotor to changes in fuel flow without considering the dynamic response of combustion chamber pressure. The model can readily be extended to include a pressure term by recognizing the possible contribution of combustion pressure to the generation of engine torque and the balance of flows into and out of the combustion chamber. The model equations are now expressed as follows.

C.1.2 Rotor Dynamics with Pressure Term

The torque expression becomes

$$Q = f(N, w_{\text{Fe}}, p_c). \tag{C.8}$$

Expanding this function as a Taylor Series yields

$$\Delta Q = \frac{\partial Q}{\partial N} \Delta N + \frac{\partial Q}{\partial w_{\text{Fe}}} \Delta w_{\text{Fe}} + \frac{\partial Q}{\partial p_c} \Delta p_c \tag{C.9}$$

Again from the acceleration of the rotor shaft, we obtain the energy equation as

$$J_R \frac{d\Delta N}{dt} = \Delta Q$$

$$= \frac{\partial Q}{\partial N} \Delta N + \frac{\partial Q}{\partial w_{\text{Fe}}} \Delta w_{\text{Fe}} + \frac{\partial Q}{\partial p_c} \Delta p_c. \tag{C.10}$$

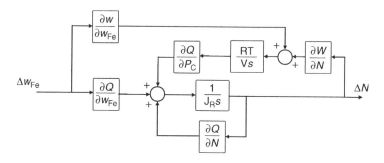

Figure C.2 Linear dynamics of rotor and pressure dynamics.

C.1.3 Pressure Dynamics

In a manner analogous to the rotor torque equation, we can observe that the airflow through the engine can be expressed as:

$$w = f(N, w_{\text{Fe}}).$$ (C.11)

As with torque, this expression can be expanded into a Taylor Series as:

$$\Delta w = \frac{\partial w}{\partial N}\Delta N + \frac{\partial w}{\partial w_{\text{Fe}}}\Delta w_{\text{Fe}}.$$ (C.12)

By considering the continuity of mass in the combustion chamber, we can express the change in combustor pressure as:

$$\frac{d(\Delta p_c)}{dt} = \frac{RT_2}{V}\Delta w.$$ (C.13)

By combining Equations C12 and C13, we obtain the following expression for the rate of change of pressure within the combustion chamber:

$$\frac{V}{RT_2}\frac{d\Delta p_c}{dt} = \frac{\partial w}{\partial N}\Delta N + \frac{\partial w}{\partial w_{\text{Fe}}}\Delta w_{\text{Fe}}.$$ (C.14)

Equations C10 and C14 can be expressed in block diagram form, as shown in Figure C.2.

C.2 Full-range Model: Extended Linear Approach

The linear approximation of engine dynamic behavior described in the previous section is valid at only a single operating point within the speed range of the engine. The various particle derivatives which form the coefficients are thus evaluated at a specific rotor speed and deviations of only a few percent are sufficient to place results in question. A study which is intent upon examining the dynamic stability of a control concept would require separate evaluations each being conducted at a specific rotor speed until the entire operating envelope has been explored.

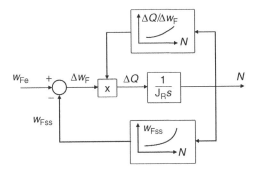

Figure C.3   Full-range model of the gas generator rotor dynamics.

The small perturbation method can be extended to allow a reasonably accurate representation of full-range throttle transients. Figure C.3 shows how this is accomplished for the gas generator rotor of a single-spool engine. By comparing the steady-state fuel flow with the actual fuel flow to the engine to develop an over- or under-fueling term Δw_f, the correct value of engine gain K_e is ensured for the full power range. Similarly, the incremental rotor torque ΔQ is arranged to vary the loop gain such that the value of the engine time constant τ_e matches the small perturbation values.

The figure shown represents one specific air inlet condition; however, it is easy to non-dimensionalize the variables as before to facilitate air inlet condition changes.

It should be recognized that this full-range modeling technique is only valid for small deviations about the engine steady running line; large fuel flow excursions will not provide representative engine behavior. Fortunately, the acceleration and deceleration limits provided by a representative model of the fuel control will tend to maintain engine response within valid limits.

C.3   Component-based Thermodynamic Models

A dynamic model of a gas turbine can be created based on descriptions of the major components. The basic notion of such a model is derived from the design process whereby individual design teams produce working designs/prototypes of the major components and then undertake the task of 'matching' these components to obtain a working engine.

The above-described design effort always results in a complete description of a component as it might be tested on an individual basis. For example, a compressor would be described as a set of performance curves as shown in Figure C.4. In the above figure, the following expressions apply:

$\frac{P_2}{P_1}$ = the overall pressure ratio;

$\frac{w\sqrt{T_1}}{P_1}$ = the non-dimensional flow rate;

$\frac{N}{\sqrt{T_1}}$ = the non-dimensional rotor speed; and

η = the isentropic efficiency.

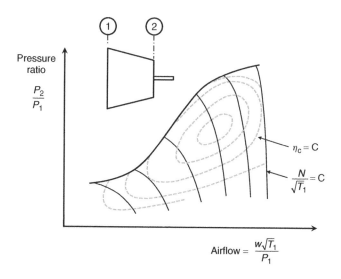

Figure C.4 Typical compressor map.

Such a map will typically have been obtained either by calculations or by rig tests. Although it will usually be a multistage compressor, only its overall performance is captured in this plot.

In this presentation, the component is represented as a one-dimensional flow device whose aerodynamic behavior is so fast that it can be regarded as quasi-steady state. Furthermore, any variations of the specific heat of air are either ignored or they are embedded in the definition of isentropic efficiency which is expressed as follows:

$$\eta_c = \left[1 - \left(P_2 / P_1 \right)^{\frac{\gamma-1}{\gamma}} \right] / \left[\frac{(T_2 - T_1)}{T_1} \right] \tag{C.15}$$

where γ is the ratio of specific heats.

Finally, this one-dimensional treatment ignores variations of performance due to inlet flow distortions and all related dynamic effects within the compressor. Despite these restrictions, it provides an excellent estimate of the migration of the compressor operating point due to changes in the throttle position or any repositioning of variable geometry such as variable nozzles.

A model of the overall engine can be constructed by applying the laws of conservation of mass and energy between the major components, which comprise:

- inlet;
- compressor;
- combustor;
- turbine;
- jet pipe;
- nozzle; and
- rotor.

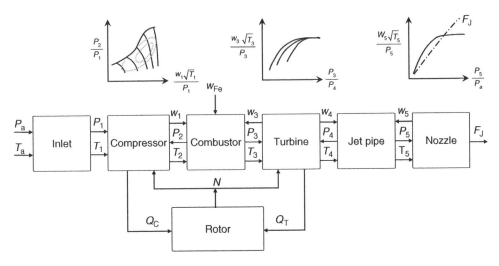

Figure C.5 Model block diagram for a single-spool turbojet engine.

Where each of the major components is described by its one-dimensional, quasi-steady performance map, an overall block diagram of a single-spool jet engine model is shown in Figure C.5. The equations required to support such a model are described in the following sections.

C.3.1   Inlet

The inlet for most modeling purposes can be considered as ideal. Such an assumption implies that the primary interest in the model is the engine performance and that any supporting tests would be conducted in a well-designed environment with a properly designed bell-mouth inlet.

We can therefore assume that:

$$P_1 = P_a$$
$$T_1 = T_a$$

where P_1 is total pressure at the face of the compressor; T_1 is the total temperature at the face of the compressor, and P_a, T_a are the ambient conditions of pressure and temperature, respectively.

For conditions other than ideal, a model of the inlet losses is required. For example, if the model is intended to describe flight conditions, it would be necessary to define a flight Mach number and assign the inlet an isentropic efficiency. In this case, the conditions at the face of the compressor would become:

$$P_1 = P_a \left[1 + \eta_i \frac{\gamma - 1}{2} M_a^2 \right]^{\frac{\gamma}{\gamma - 1}} \tag{C.16}$$

$$T_1 = T_a \left[1 + \frac{\gamma - 1}{2} M_a^2 \right]$$
(C.17)

where M_a is flight Mach Number, η_i is inlet efficiency, and γ is the ratio of specific heat for air ($= 1.4$).

The isentropic efficiency of an inlet can be expected to be quite high (typically about 95%).

C.3.2 Compressor

The model of a compressor is based on the performance maps as described earlier. The inputs to this block are:

N: rotor speed;
T_1: inlet total temperature;
P_1: inlet total pressure; and
P_2: compressor discharge pressure.

With these inputs, it is possible to access the performance maps and obtain values of non-dimensional airflow $w_1 \sqrt{T_1}/P_1$ and efficiency η_c.

Some convenient means of representing these maps is required. The most common method is the direct use of the data in tabular form and some form of table look-up algorithm. Once the map data has been obtained, the outlet conditions can be computed as follows:

$$T_2 = T_1 + \frac{T_1}{\eta_c} \left[1 - \left(\frac{P_2}{P_1} \right)^{\frac{\gamma - 1}{\gamma}} \right]$$
(C.18)

and

$$w_2 = \frac{w_1 \sqrt{T_1}}{P_1} \frac{P_1}{\sqrt{T_1}}.$$
(C.19)

With these data, the torque required to drive the compressor is calculated from:

$$Q_c = \frac{60 C_P J w_1 (T_2 - T_1)}{2\pi N}$$
(C.20)

where Q_c is the compressor torque, J is the mechanical equivalent of heat, and C_P is the average specific heat of air through the compressor.

C.3.3 Combustor

For the purposes of modeling, the combustor is treated as an accumulator where the equation of continuity of mass determines the changes in pressure within the combustor. In addition, it is treated as a chemical reactor whereby fuel is burned in the presence of air to raise the temperature of the outlet gas.

The inputs to this block (per Figure C.5) are:

w_1: compressor air flow;

w_3: turbine gas flow;

T_2: compressor delivery temperature; and

w_{Fe}: engine fuel flow rate.

A simplified version of the law of continuity of mass is commonly used to describe the rise in pressure within the combustor:

$$\frac{dP_2}{dt} = \frac{RT_2}{V}(w_2 + w_{Fe} - w_3) \tag{C.21}$$

where R is the gas constant, V is the combustor volume, and P_2 is the compressor delivery pressure.

Integration of this expression with respect to time provides an estimate of the dynamic behavior of combustor pressure. The combustor outlet pressure can be calculated as a simple proportionality from:

$$\frac{P_2 - P_3}{P_2} = \text{PLF} \tag{C.22}$$

where PLF is the combustor pressure loss factor. Values of PLF are of the order 0.04–0.05; however, the estimate of pressure loss can be improved somewhat by making it a function of non-dimensional flow rate:

$$\frac{P_2 - P_3}{P_2} = \text{PLF} \times \frac{R}{2}\left(\frac{w\sqrt{T_2}}{AP_2}\right)^2 \tag{C.23}$$

where the pressure loss factor is now expressed as:

$$\text{PLF} = k_1 + k_2\left(\frac{T_3}{T_2} - 1\right) \tag{C.24}$$

and $w\sqrt{T_2}/AP_2$ is the non-dimensional flow at an average cross-section of the combustor.

The latter expression for pressure loss will provide improved accuracy across the range of engine operation; however, many models used for system analysis use the simpler form with good results.

Finally, a simple expression for the change in gas temperature across the combustor is obtained from the equation for conservation of energy:

$$T_3 - T_2 = \frac{w_{Fe}\Delta H_{Fe}}{C_{Pa}w_2} \tag{C.25}$$

where ΔH_{Fe} is the lower calorific value of fuel and C_{Pa} is the average specific heat across the combustor.

Again, a more rigorous treatment of this equation would account for the variation with temperature of the lower calorific value of fuel and the values of specific heat at the entrance and exit of the combustion chamber (expressed as enthalpies). However, this addition complexity will contribute little to the system analysis unless the model is also being employed to estimate engine thermodynamic performance.

C.3.4 *Turbine*

The model of the turbine is based on the performance maps as described earlier. The inputs to the turbine block in Figure C.5 are rotor speed N, turbine inlet temperature T_3, turbine inlet pressure P_3, and turbine outlet pressure P_4.

With these inputs, it is possible to access the performance maps and obtain values of non-dimensional flow $w_3\sqrt{T_3}/P_3$ and turbine efficiency η_T.

As with the compressor maps, some convenient means of representing the map data is required. Once the map data has been obtained, the turbine outlet conditions can be computed as follows:

$$\frac{T_3 - T_4}{T_3} = \eta_T \left[1 - \left(\frac{1}{P_3/P_4} \right)^{\frac{\gamma-1}{\gamma}} \right] \tag{C.26}$$

and

$$w_3 = \frac{w_3\sqrt{T_3}}{P_3} \left(\frac{P_3}{T_3} \right). \tag{C.27}$$

Similarly, the turbine torque can be calculated as

$$Q_T = \frac{60 C_{PT} J w_3 (T_3 - T_4)}{2\pi N} \tag{C.28}$$

where Q_T is the turbine torque and c_{PT} is the average specific heat of the gas through the turbine.

The block on turbine modeling would not be complete without a comment on the applicability of the turbine performance maps to this calculation. A full performance map for a typical turbine is shown in Figure C.6. Basically, the nozzles of the turbine behave

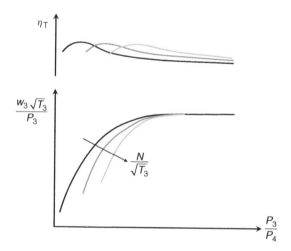

Figure C.6   Typical turbine performance characteristics.

to a very large extent like any nozzle. This suggests that the flow characteristic is largely independent of rotor speed.

While largely true, there is some observable dependence on rotor speed. This is because some of the expansion that takes place within the rotor. Furthermore, if the turbine is multistage we would expect to see increasing speed dependence in the data with the choking value of $w_3 \sqrt{T_3}/P_3$ being reduced.

The efficiency curves in the figure show a somewhat more pronounced speed dependency; however, calculations will show that the compressor operating line will largely depend on the flow characteristic while the turbine inlet temperature will show greater dependence on efficiency.

The availability of a turbine map for the target engine model is much less an issue than the compressor. If actual data is available, it should obviously be used; however, very good results can be obtained by using published data from another similar turbine and scaling the data about the design point to fit the purpose at hand. Indeed, quite adequate results for system analysis can be obtained by treating the turbine as a nozzle with little or no speed dependence.

The modeler is reminded that most of the system analysis is aimed at obtaining good dynamic response while avoiding compressor surge. The effect of variations in the turbine data on this result is typically not very pronounced.

Obviously, if the model is intended for use in estimating the thermodynamic performance of the engine which is to be forwarded to turbine and compressor designers as part of an engine design and development project, then the accuracy of the model is much more important. Under these circumstances, however, the modeler will usually have access to the more accurate data provided to him by the turbine designer.

C.3.5 Jet Pipe

The model of the jet pipe is in all respects identical to the combustor. In the example depicted in Figure C.5 there is no afterburner included and the model reduces to a simple accumulator with a pressure loss. The equations for the jet pipe are therefore:

$$\frac{dP_4}{dt} = \frac{RT_4}{V_j}(w_4 - w_5) \tag{C.29}$$

where P_4 is the turbine exit pressure, T_4 is the turbine exit temperature, V_j is the jet pipe volume, w_4 is the turbine flow, and w_5 is the nozzle flow.

We can use the same pressure loss factor approach used in the combustor analysis, that is:

$$\frac{P_4 - P_5}{P_4} = \text{PLF.} \tag{C.30}$$

This is typically taken as a constant or as a function of non-dimensional flow in a manner that is exactly analogous to the combustor.

Similarly, if the engine is fitted with an afterburner, the temperature rise is calculated using the law of conservation of energy in the same manner as the description of this phenomenon for the combustor.

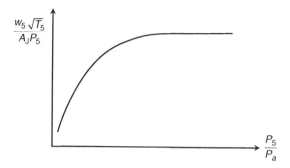

Figure C.7   Nozzle flow characteristic.

C.3.6 Nozzle

For modeling purposes, the nozzle is treated as an aerodynamic device which is described in non-dimensional form (Figure C.7). The data in Figure C.7 can be obtained from any text book on compressible flow [2].

The nozzle area A_j is included in the data shown. This treatment allows for the convenient control of the nozzle area to adjust performance or allow for variable nozzle control, as would be necessary in the use of an afterburner.

C.3.7 Rotor

The final block of the diagram shown in Figure C.4 is the engine rotor. The inputs to this block are turbine torque Q_T and compressor torque Q_C.

The equation governing the speed of the rotor is conservation of energy in the form:

$$\frac{dN}{dt} = \frac{60}{2\pi J_R}(Q_T - Q_C) \tag{C.31}$$

where J_R is the rotor polar moment of inertia and N is the rotational speed in rpm.

References

1. Schwarzenbach, J. and Gill, K. (1992) *System Modeling and Control A*, 3rd edn, Butterworth, Heinemann.
2. Shapiro, A.H. (1953) *The Dynamics and Thermodynamics of Compressible Flow*, Ronald Press.

Appendix D

Introduction to Classical Feedback Control

This appendix provides an introduction to classical feedback control covering the basic principles of stability and control as it applies to single-input single-output linear control systems. At the heart of the discussion is the concept of the transfer function as a means to represent the dynamic elements of a control system in block diagram form as an alternative to the use of traditional mathematical differential equations [1].

Also contained in the discussion that follows is the application of Laplace transforms to observe control system behavior in the 'frequency domain', providing the system designer with a unique graphical view of control system behavior as control system parameters are varied.

D.1 Closing the Loop

The generic closed-loop control system (see Figure D.1) can be described as a means to control the output of a process by comparing what is required with the actual output, and using the output from this comparison to generate controlling actions that change the process output toward that required.

In the diagram, the input to the system (that required) is compared to the feedback, which is a measure of the actual output, to establish the 'error'. This error is used by the controller to generate an input to the process that causes the output to move toward the required setting. Also shown in the diagram is a disturbance input representing external changes from outside the control loop that may affect the process.

Examples of disturbances in a gas turbine propulsion application are changes in inlet conditions or changes in mechanical power extraction due to demands from either the electrical generation system or the hydraulic system. Any changes in the process output due to external disturbances are fed back to the controller, which will provide the appropriate corrective action.

It can intuitively be seen that for the error to be very small the controller must be 'sensitive'; that is, small errors must be capable of generating significant response if the

Gas Turbine Propulsion Systems, First Edition. Bernie MacIsaac and Roy Langton.
© 2011 John Wiley & Sons, Ltd. Published 2011 by John Wiley & Sons, Ltd.

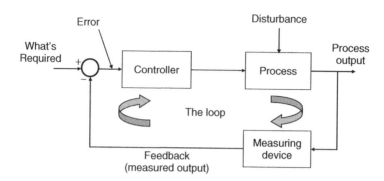

Figure D.1   The generic closed-loop control system.

control of the process is to be effective. In other words, the controller must have a 'high gain'. Not so intuitive is the fact that, with high-gain closed-loop systems, time delays around the loop can cause the system to be 'unstable'.

Understanding the concept of 'stability' and having the tools to design closed-loop systems that are well behaved from a stability and performance perspective is what classical feedback control is all about. The intention of this appendix is to provide the reader with a brief insight into the subject.

D.2   Block Diagrams and Transfer Functions

First we need to understand how to represent traditional time-differential equations in block diagram form, from which it is much easier to visualize dynamic behavior than using pure mathematical concepts. To best explain this concept, consider the simple spring-mass system shown in Figure D.2.

The force-balance equation for this system is:

$$(x_i - x_o)K - f\left[\frac{dx_o}{dt}\right] = M\left[\frac{d^2 x_o}{dt^2}\right]. \tag{D.1}$$

Using the D notation (where $D = d/dt$) we can rewrite this equation as:

$$(x_i - x_o)K - fDx_o = MD^2 x_o. \tag{D.2}$$

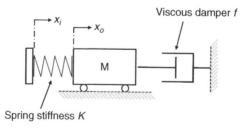

Figure D.2   Spring-mass system.

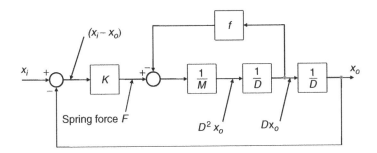

Figure D.3   Spring-mass system block diagram.

In order to represent this system in block diagram form, we put the highest derivative term on the left-hand side of the equation:

$$D^2 x_0 = \frac{1}{M}[(x_i - x_0) - f D x_0].$$ (D.3)

It is now an easy task to construct a block diagram, as shown in Figure D.3.

The output from each block is the product of the input to that block and the contents of the block. The block containing the $1/D$ terms represents the integration process, with the constant of integration being the initial value of the integrator output at $t=0$.

The diagram of Figure D.3 shows a system with two feedback loops. The inner loop feeds back the viscous damper force, which is subtracted from the spring force to yield the net force acting on the mass. Mass position is then fed back and subtracted from the input to obtain the spring deflection.

For any closed-loop system with negative feedback, the relationship between the input and output of the system can be determined from:

$$\frac{\text{Output}}{\text{Input}} = \frac{\text{Product of forward path elements}}{1 + \text{Product of all elements around the loop}}.$$ (D.4)

We can therefore rationalize the inner loop as a single expression relating the velocity of the mass and the spring force as:

$$\frac{D x_0}{F} = \frac{1}{(MD + f)}.$$ (D.5)

The right-hand side of the above equation is called the transfer function of the inner loop and we can now replace the inner loop in Figure D.3 by a single block which becomes an element in the forward path of the overall system. The closed-loop transfer function for the complete system can therefore be developed as:

$$\frac{x_0}{x_i} = \frac{(K/D)(1/(MD + f))}{1 + (K/D)(1/(MD + f))}$$ (D.6)

which can be further rewritten as:

$$\frac{x_0}{x_i} = \frac{1}{(D^2/\omega_n^2) + (2\zeta/\omega_n)D + 1}.$$ (D.7)

This is the standard method of expressing a second-order transfer function where the natural frequency ω_n is equal to the spring stiffness K divided by the mass M. The term ζ is referred to as the damping ratio as is equal to $f\omega_n/2K$.

When the damping ratio is unity, the roots of the second-order expression are real and identical. Further increases in the damping ratio cause the real roots to separate, with the separation becoming greater as the damping ratio increases. This regime of solutions represents the state where the viscous damping dominates and the output response to changes in input is sluggish.

As damping ratio is reduced below 1.0, the roots become two complex conjugates. For values below about 0.7, the output response to changes in the input becomes oscillatory to the point where, for a damping ratio of zero, we will have sustained oscillations with a frequency equal to the natural frequency ω_n.

The importance of the above discussion will become clearer later in Section D.5 where Laplace transforms and root locus techniques are described.

D.3   The Concept of Stability

Stability is a qualitative description of the performance of the system. Ideally we would like the output to respond quickly and precisely to changes in the input. However, if we continue to increase the gain (sensitivity) of the controller in an attempt to improve response and accuracy, the output will eventually begin to exhibit oscillatory behavior. This phenomenon is caused by time delays around the control loop and can lead to instability. When instability is reached, the output may oscillate continuously or the oscillations may continue to increase until the output reaches its maximum limit.

To understand this effect, let us return to the generic closed-loop example in the previous section with a few changes:

1. insert a switch in the error signal line;
2. assume that the input to the control loop is held constant; and
3. assume that the disturbance input is zero.

Now consider a sinusoidal input to the controller with the switch in the error path set open as indicated in Figure D.4. Note that the controller is now represented by a simple gain G.

From the diagram it can be seen that if the feedback signal (measured output) is equal in magnitude and exactly 180° out of phase with the originating signal, then the oscillations will continue if the switch is closed even with a fixed input to the control loop. In other words, the oscillation just continues at the same amplitude. When this condition occurs, the system is said to have 'marginal stability'.

D.3.1   The Rule for Stability

When the phase lag around the loop is 180° (1/2 a cycle) in a closed-loop system (with negative feedback), the gain around the loop must be less than 1.0 for the system to be stable. Note also that if the loop gain is greater than unity when the phased shift around the loop is 180°, the system will exhibit divergent oscillations.

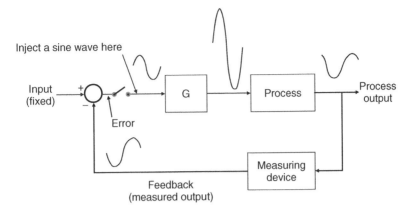

Figure D.4   Illustration of the concept of stability.

From the above rule it follows that, in order to determine the stability of a closed-loop system, we need to analyze the open-loop behavior of the system. The following section describes the techniques employed either by analysis or testing to determine the acceptability of closed-loop control systems from a response and stability perspective.

D.4   Frequency Response

Since closed-loop systems exhibit oscillatory behavior when approaching instability, the concept of frequency response as a method of both analysis and test of closed-loop systems has become perhaps the most common technique used by the control industry today to assess the relative stability of these systems.

Using either analytical methods or by testing, the objective is to inject sinusoidal inputs over the frequency range of interest and to measure the amplitude ratio and phase angle shift between input and output at each frequency. For linear systems, the results are independent of signal amplitude.

Typically, physical systems follow the input command closely at low frequencies with only small phase lags. As the frequency goes beyond the bandwidth of the system, however, the results show rapidly increasing attenuation and phase lag.

As mentioned above it is the open-loop elements that determine the stability of a closed-loop system. The analysis (or testing) should therefore focus on the open-loop frequency response characteristics for the purposes of determining the relative stability of the system.

D.4.1   Calculating Frequency Response

Mathematically, frequency response is expressed as the ratio of the output vector to the input vector as a function of $j\omega$ as follows:

$$\frac{x_o}{x_i}(j\omega) = \left| \frac{x_o}{x_i} \right| \angle \frac{x_o}{x_i}. \qquad (D.8)$$

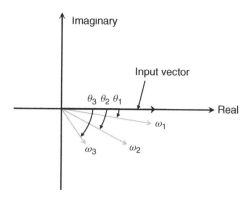

Figure D.5 Frequency response vectors in the complex plane.

The right-hand side of the above equation represents the ratio of the output and input vector moduli and the phase angle between the two vectors.

The process for calculating the frequency response is completed by substituting $D = j\omega$ in the transfer functions of the system for a range of values of ω. Figure D.5 shows the relationship between the input and output vectors in the complex plane for three different frequencies.

The input vector remains constant for all frequencies and the output vectors show attenuation and phase shift relative to the input vector. It can be seen from the figure that the length of the response vector can be obtained from:

$$|x_o| = \sqrt{(R^2 + I_m^2)} \qquad (D.9)$$

where R and I_m are the real and imaginary components of the vector. The length of this vector relative to the input vector is the amplitude ratio. Similarly, the phase shift between the input and output vectors is obtained from:

$$\text{Phase shift } (\theta) = \tan^{-1}\left(\frac{I_m}{R}\right). \qquad (D.10)$$

The following are examples of frequency responses for two of the most common linear transfer functions. The first-order lag is of the form:

$$\frac{1}{(1 + \tau D)}$$

where τ is referred to as the time constant with units of time. The second example is the second-order system with a transfer function identical to that described in Section A.1.2 above and having the form:

$$\frac{1}{(D^2/\omega_n^2) + (2\zeta/\omega_n)D + 1}$$

where ω_n is the natural frequency and ζ is the damping ratio.

First we need to address conventions associated with frequency response plots. For convenience, both amplitude ratio and frequency are plotted in logarithmic form. Amplitude ratio is converted into decibels (dB) according to the formula:

$$\text{Gain (dB)} = 20\log_{10} \text{(amplitude ratio)}. \qquad (D.11)$$

Using the above conventions makes it relatively easy to generate frequency response gain plots since they have linear asymptotes and transfer functions in series can be multiplied together by simply adding their gains.

Figures D.6 and D.7 show frequency response plots for the above two examples.

The first-order lag plot has a frequency scale normalized as a function of the time constant of the lag. Note that the two gain asymptotes intersect at a frequency of $1/\tau$. Below this frequency, the gain is asymptotic to 0 dB (which is equivalent to an amplitude ratio of 1.0). Above this frequency, the gain reduces by 6 dB (a factor of 2.0) for each doubling of the frequency (an octave). The phase lag curve goes from 0 to 90° passing through 45° at the break frequency $1/\tau$.

The second-order system plot has a frequency scale as a fraction of the natural frequency ω_n. The gain curves are asymptotic about the natural frequency with the gains at frequencies below this point tending toward 0 dB. Above this frequency, the asymptote shows an attenuation of 12 dB per octave (twice that of the first-order lag). As shown, when damping ratios are below about 0.7 there is a magnification of the input. When the damping ratio is a low as 0.05, this magnification is +20 dB which is equivalent to an amplitude ratio of 10.0.

The phase lag curves pass through 90° of phase lag at the natural frequency on their way to a maximum of 180°. The shape of the phase curves changes with the damping ratio.

The above examples cover all that is needed to generate frequency response plots, since all transfer functions are reducible to a number of first-order and second-order terms that

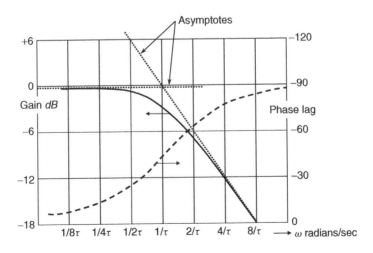

Figure D.6 First-order lag frequency response.

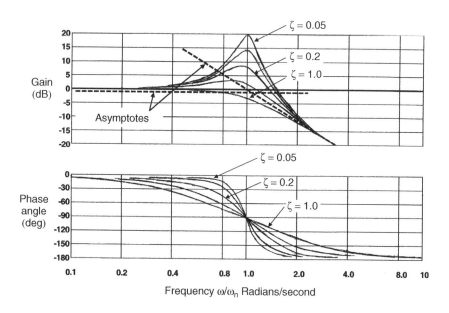

Figure D.7   Second-order system frequency response.

can be treated as described above. Gains and phase shifts are simply added to obtain the composite gains and phase angles for the complete system.

For stability analysis it is the open-loop frequency response which is of interest. Plots of this type are called 'Bode diagrams' and an example is shown in Figure D.8. This example has an arbitrary third-order open-loop transfer function comprising an integrator and two first-order lags with different time constants.

The important feature of the Bode diagram is that it shows the 'stability margins' of the system and thereby indicates the type of behavior the system would exhibit when the loop is closed. The definitions of the stability margins are as follows.

- **Gain margin:** The gain increase that would result in the gain curve crossing the 0 dB line at the frequency corresponding to 180° of open-loop phase lag.
- **Phase margin:** The additional phase lag that would result in 180° of open-loop phase at a frequency corresponding to 0 dB of open-loop gain.

In the example, the gain and phase margins are approximately 18 dB and 62°, respectively.

The appropriate stability margins depend upon the system, its application, and the confidence that can be applied to the parameters used in the analysis. Some systems are more gain sensitive than phase sensitive and vice versa. In our example, gain is the more critical stability criterion and increasing the gain by 18 dB (which is about a factor of 8) would result in marginal stability with sustained oscillations at about 7 rad/s.

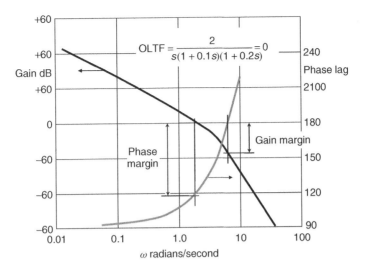

Figure D.8 Bode diagram example.

D.5   Laplace Transforms

Laplace transforms provide the control engineer with an alternative way to visualize the behavior of closed-loop control systems through the introduction of a simple transform methodology, whereby the process of developing the time response of systems to various time functions including step, ramp, and frequency inputs is reduced to simple algebraic mathematics.

The Laplace transform methodology can be considered to be similar to the use of logarithms, where the processes of multiplication and exponent functions are simplified using tables to transform the original function into a new domain. After performing a simplified process (e.g., addition for multiplication), antilogarithms are used to convert the answer from the transformed domain back to the real world.

The purpose of the Laplace transform is to transform functions of time into the frequency domain (also known as the s-plane) where:

$$F(s) = \mathcal{L}f(t) \tag{D.12}$$

where the term $\mathcal{L}$ denotes Laplace transform and

$$\mathcal{L}f(t) = \int_0^\infty f(t)e^{-st}\, dt. \tag{D.13}$$

In the s-plane, $s = \alpha + j\omega$ where the imaginary term represents frequency and the real term determines the rate of decay (or expansion) of that frequency, as indicated on Figure D.9 which illustrates the concept for six different locations in the plane.

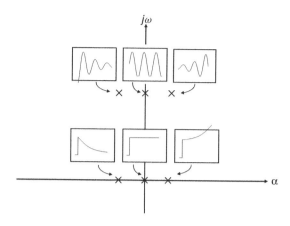

Figure D.9   The frequency domain (s-plane).

Laplace transforms allow us to convert functions of time, including transfer functions of the dynamic elements of the system, into the frequency domain. Standard tables are available showing the Laplace transforms of many common time functions such as the step, ramp, and sine functions. Conversion of transfer functions simply requires the substitution of the D operator with the Laplace operator s.

Calculating system response involves simple algebra to generate the system output, which is the product of the input transform and the system transform. Since the system output is still in the Laplace domain, we use the inverse Laplace process (again using standard tables) to convert the answer back into the time domain.

An example of this process is shown below. Consider a step input to a first-order lag beginning at $t = 0$. Figure D.10 represents this diagrammatically for both the time domain and the s domain. The term $H(t)$ represents the step function which equals zero for $t \leq 0$ and 1.0 for $t \geq 0$.

The Laplace transform for this step function is $1/s$; therefore the response in the s domain is:

$$x(s) = \frac{1}{s(1 + \tau s)} = \frac{1}{s} - \frac{\tau}{(1 + \tau s)} \text{(by partial fractions)}. \qquad (D.14)$$

This can be written in the following form:

$$x(s) = \frac{1}{s} - \frac{1}{[s + (1/\tau)]} \qquad (D.15)$$

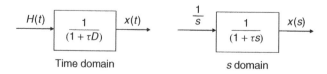

Figure D.10   Step response diagram for time and s domains.

If we now convert these two terms back into the time domain, we obtain:

$$x(t) = 1 - e^{-t/T} \text{ which is valid for } t \geq 0. \tag{D.16}$$

The above is a very brief overview of the application of Laplace transforms to linear control systems. Fortunately, much of the previous procedures discussed for the D operator remain essentially unchanged. For example, the calculation of frequency response is the same except that we now substitute $s = j\omega$ in our transfer functions and proceed as before. In control systems practice, the terms s and D are often used interchangeably.

There is an additional and very powerful process that remains to be discussed. This involves Laplace transforms and the frequency domain, which allows the control system engineer to visualize closed-loop behavior from the examination of the open-loop roots. This process is called 'root locus' and is described in the following section.

D.5.1 Root Locus

The root locus technique is developed from the characteristic equation for any closed-loop system which is:

$$1 + \text{the open-loop transfer function} = 0. \tag{D.17}$$

The roots of this equation define the dynamic characteristics of the system. It also defines the condition for marginal stability, where sustained oscillations will occur if gain around the loop is unity and the phase lag around the loop is $180°$. The characteristic equation can also be expressed as:

$$\text{Product of all the elements around the loop} = -1 = |1.0|\angle 180°. \tag{D.18}$$

Let us consider a typical control system example having a characteristic equation of the form:

$$\frac{K(s + z_1)(s + z_2)}{(s + p_1)(s + p_2)(s + p_3)} = |1.0|\angle 180°. \tag{D.19}$$

The z values in the numerator are called zeros since the expression goes to zero when s is set to any of the z values. In the denominator, the p values are referred to as poles since the expression goes to infinity when s is set to any of the p values.

It follows that, if we can define a locus of all of the points in the frequency domain (the s-plane) in which the sum of all of the vector angles from all of the zeros to any point on the locus minus the sum of all of the vector angles from all of the poles to the same point on the locus is equal to $180°$, we will be able to how see the closed-loop roots of the system move in the s-plane as the value of K varies from zero to infinity.

The value of K at any point on the locus is simply the product of the vector lengths (moduli) from the zeros to the point on the locus divided by the product of the vector lengths from the poles to the same point.

The location of these root loci in the s-plane is a simple process supported by a number of simple rules that allow the analyst to quickly sketch these loci. The number of loci equals the number of poles and these loci travel from the poles to the zeros or, in the case where there are no zeros, from the poles to infinity along predetermined asymptotes.

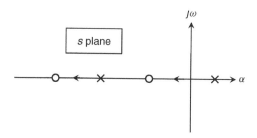

Figure D.11   Root loci with real axis poles and zeros.

Below is a summary of the rules for root locus construction, followed by an example using the open-loop system transfer function that was used to generate the Bode diagram of Figure D.8.

D.5.2   Root Locus Construction Rules

Root loci can be constructed easily using a few simple rules that can be applied to any linear control system. These rules are summarized below and are fairly intuitive to the first-time user.

Rule 1: Root loci travel from each pole, where the loop gain is zero, to either an associated zero or to infinity along a predetermined asymptote.

Rule 2: A locus will always be present along the real axis of the s-plane to the left of an odd number of poles plus zeros (see Figure D.11).

Rule 3: When there are more poles than zeros (most cases), there are additional zeros located at infinity. These loci will be asymptotic to straight lines with angles defined by:

$$\theta = \frac{180 + i360}{n - m}$$

where n and m are the number of zeros and poles, respectively. This equation applies for $i = \pm 0, \pm 1, \pm 2$ and so on until all $(n - m)$ angles not differing by multiples of $360°$ are obtained.

Rule 4: The starting point on the real axis from which the asymptotes radiate (point r of Figure D.12) is given by the equation:

$$\alpha = \frac{\Sigma(\text{pole values}) - \Sigma(\text{zero values})}{n - m}.$$

Rule 5: When two adjacent poles lie on the real axis, there will be a breakaway point on the locus between the two poles as shown in the example of Figure D.13.

In order to illustrate how root locus theory works, we need to connect this technique with the conventional analysis methods outlined earlier. To accomplish this, consider the

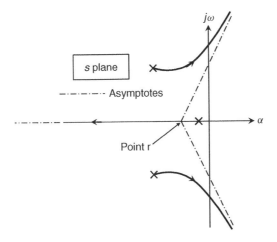

Figure D.12 Root loci asymptotes for a three-pole system.

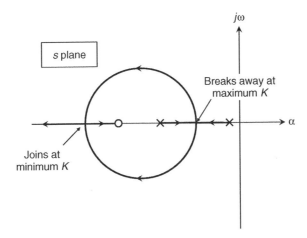

Figure D.13 Breakaway point example.

following example which uses the same open-loop transfer function used in the Bode diagram of Figure D.8. This is shown below replacing the D operator with the Laplace operator s and expressing the loop gain as the variable K:

$$\text{Open-loop transfer function} = \frac{K}{s(1 + 0.1s)(1 + 0.2s)}.$$

Rearranging and expressing the system characteristic equation as $1 +$ the loop $= 0$ yields:

$$1 + \frac{50K}{s(s + 10)(s + 5)} = 0$$

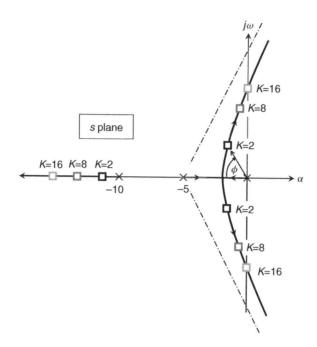

Figure D.14   Root locus plot showing closed-loop roots.

We can now represent the above system in the frequency domain as an open-loop system with three poles along the real axis: one at the origin, one at -5 and one at -10.

Based on the above rules we can construct the root loci to show how the closed-loop roots move as the loop gain is increased from zero to infinity. Figure D.14 shows this system, its open-loop poles, the associated root loci, and the closed-loop roots for three different values of K.

As the value of K increases, the closed-loop roots move along the loci. The locus between the two open-loop poles divides into two separate loci which then move toward the asymptotes, finally crossing the $j\omega$ axis at a value of $K = 16$. For this value of loop gain, the system is marginally stable having sustained oscillations of about 7 rad/s. This is exactly the same prediction that we obtained from the Bode diagram of Figure D.8.

The closed-loop roots along these loci are complex conjugate pairs and the natural frequency of this second-order element is equal to the length of the vector from the origin to the closed-loop root, as indicated by the arrow in the diagram. The damping ratio associated with the complex conjugate roots is equal to the cosine of the angle ϕ shown in the figure for the root at $K = 2$. It can therefore be seen how the standard form used for second-order systems is useful in visualizing how the system behaves dynamically.

The third real root moves to the left along the real axis toward minus infinity as the gain is increased.

The above simple example demonstrates how simple it is to visualize the closed-loop behavior of a system by representing the open-loop characteristics in the frequency domain. The concept of frequency response is also illustrated by this technique since the

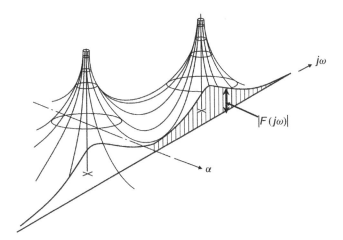

Figure D.15   3-D interpretation of a second-order system.

amplitude ratio is simply the product of vectors from the poles to any point on the $j\omega$ axis, and the phased shift is the sum of the angles from the same vectors.

Figure D.15 is a three-dimensional interpretation of a second-order system with two complex conjugate poles in the s–plane, in an attempt to assist the reader in visualizing the above concept. By taking a slice through the $j\omega$ axis, the shape of the contour is equivalent to the amplitude ratio of the frequency response.

Reference

1. Langton, R. (2006) *Stability and Control of Aircraft Systems*, John Wiley & Sons, Ltd, UK.

Index

Gas Turbine Propulsion Systems, First Edition. Bernie MacIsaac and Roy Langton.
© 2011 John Wiley & Sons, Ltd. Published 2011 by John Wiley & Sons, Ltd.

Printed and bound by CPI Group (UK) Ltd, Croydon, CR0 4YY

Printed and bound by CPI Group (UK) Ltd, Croydon, CR0 4YY

16/04/2025

14658831-0003